This series aims to report new developments in mathematical research and teaching – quickly, informally and at a high level. The type of material considered for publication includes:

1. Preliminary drafts of original papers and monographs
2. Lectures on a new field, or presenting a new angle on a classical field
3. Seminar work-outs
4. Reports of meetings

Texts which are out of print but still in demand may also be considered if they fall within these categories.

The timeliness of a manuscript is more important than its form, which may be unfinished or tentative. Thus, in some instances, proofs may be merely outlined and results presented which have been or will later be published elsewhere.

Publication of *Lecture Notes* is intended as a service to the international mathematical community, in that a commercial publisher, Springer-Verlag, can offer a wider distribution to documents which would otherwise have a restricted readership. Once published and copyrighted, they can be documented in the scientific literature.

Manuscripts

Manuscripts are reproduced by a photographic process; they must therefore be typed with extreme care. Symbols not on the typewriter should be inserted by hand in indelible black ink. Corrections to the typescript should be made by sticking the amended text over the old one, or by obliterating errors with white correcting fluid. Should the text, or any part of it, have to be retyped, the author will be reimbursed upon publication of the volume. Authors receive 75 free copies.

The typescript is reduced slightly in size during reproduction; best results will not be obtained unless the text on any one page is kept within the overall limit of 18 x 26.5 cm (7 x 10 ½ inches). The publishers will be pleased to supply on request special stationery with the typing area outlined.

Manuscripts in English, German or French should be sent to Prof. Dr. A. Dold, Mathematisches Institut der Universität Heidelberg, Tiergartenstraße or Prof. Dr. B. Eckmann, Eidgenössische Technische Hochschule, Zürich.

Die „*Lecture Notes*" sollen rasch und informell, aber auf hohem Niveau, über neue Entwicklungen der mathematischen Forschung und Lehre berichten. Zur Veröffentlichung kommen:

1. Vorläufige Fassungen von Originalarbeiten und Monographien.
2. Spezielle Vorlesungen über ein neues Gebiet oder ein klassisches Gebiet in neuer Betrachtungsweise.
3. Seminarausarbeitungen.
4. Vorträge von Tagungen.

Ferner kommen auch ältere vergriffene spezielle Vorlesungen, Seminare und Berichte in Frage, wenn nach ihnen eine anhaltende Nachfrage besteht.

Die Beiträge dürfen im Interesse einer größeren Aktualität durchaus den Charakter des Unfertigen und Vorläufigen haben. Sie brauchen Beweise unter Umständen nur zu skizzieren und dürfen auch Ergebnisse enthalten, die in ähnlicher Form schon erschienen sind oder später erscheinen sollen.

Die Herausgabe der „*Lecture Notes*" Serie durch den Springer-Verlag stellt eine Dienstleistung an die mathematischen Institute dar, indem der Springer-Verlag für ausreichende Lagerhaltung sorgt und einen großen internationalen Kreis von Interessenten erfassen kann. Durch Anzeigen in Fachzeitschriften, Aufnahme in Kataloge und durch Anmeldung zum Copyright sowie durch die Versendung von Besprechungsexemplaren wird eine lückenlose Dokumentation in den wissenschaftlichen Bibliotheken ermöglicht.

Lecture Notes in Mathematics

A collection of informal reports and seminars
Edited by A. Dold, Heidelberg and B. Eckmann, Zürich

6

Hans Hermes

Universität Freiburg i. Br., Math. Institut
Abteilung für math. Logik und Grundlagen der Mathematik

Term Logic with Choice Operator

Revised and Enlarged Translation
of the first German Edition

Springer-Verlag
Berlin · Heidelberg · New York 1970

Typescript: Lieselotte Patton

First English Edition
revised and enlarged from the first German Edition
published 1965

Contents

1. Introduction. Foreword to the Revised Edition.

Whitehead and Russell [7] have been among the first to investigate a descriptive operator in the framework of mathematical logic. Hilbert and Bernays [5] have generalized this operator by introducing a choice operator ε . Cf. also Rosser [6].

A well known procedure in usual predicate logic is to start with a semantic foundation and then to give a system of rules which is both sound and complete. Henkin [3] has extended this procedure to predicate logic with choice operator, even to the theory of types.

If such a logic is built up - cf. section 3 for predicate logic (with choice opertor) - it may be remarked that there exists a certain duplicity between formulas on one side and terms on the other side. One may ask whether it is possible to deal without formulas or without terms. This indeed is the case. In this paper we introduce a kind of logic where we have only one kind of expressions called terms, and no formulas. This logic, which includes the choice operator, will be called term logic. We omit the predicate symbols and restrict ourselves to function symbols. The choice operator ε is applied to terms and generates again a term. The sentential connectives $\neg$, $\wedge$ and the identity sign $=$ are treated as function symbols. Quantifiers are not needed since they are definable by the ε - operator (cf. e. g. Hilbert - Bernays [5]: $\vee x\alpha$ may be considered as an abbreviation for the term obtained from the term α by substituting $\varepsilon x\alpha$ for x , and $\wedge x\alpha$ as an abbreviation for the term obtained from α by substituting $\varepsilon x\neg\alpha$ for x .

In order to introduce the semantical concepts we start with the notion of a semantic basis . This notion is defined in section 2 for predicate logic and in section 3 for term logic. A semantic basis for term logic is a septuple $\langle \omega, \pi, \mathfrak{F}, \mathfrak{i}, \mathfrak{n}, \mathfrak{t}, \mathfrak{a} \rangle$. Here ω is the individual domain. π is a subset of ω. The elements of π in a certain sense represent "the truth". We postulate that π and $\omega - \pi$ are not void; hence ω has at least two elements. The function symbols are interpreted by functions over ω which are elements of $\mathfrak{F}$. It is required that the special function symbols $=, \neg, \wedge$ are interpreted always by the functions $\mathfrak{i}, \mathfrak{n}, \mathfrak{t}$ belonging to the semantic basis (in this sense these function symbols are constants, the other function symbols variables). $\mathfrak{i}, \mathfrak{n}, \mathfrak{t}$ are supposed to obey certain laws which connect these functions with identity, negation, and conjunction.

Finally α is a choice operator over ω.

If π has exactly one element we have a restricted semantic basis. If only restricted bases are admitted we get the semantic for restricted term logic.

In sections 4 and 5 we show that predicate logic with choice operator (as defined in section 2) in a certain sense is embeddable in term logic, and vice versa.

In section 7 we give a system of rules for term logic (and for restricted term logic). This is a calculus using sequents. We prove that this calculus is sound (section 8) and complete (sections 9 to 15). For the completeness proof we use the procedure introduced by Henkin [2] and modified by Hasenjaeger [1].

In section 11 substitution is defined as a convenient generalization of elementary substitution.

The gist of the completeness proof is explained for term logic in section 13 and for restricted term logic in section 15.

*

The Revised Edition differs from the first essentially in the following points:

(1) It is proved that (under certain assumptions) not only predicate logic is embeddable in term logic but also that term logic is embeddable in predicate logic.

(2) The notion of restricted term logic is introduced and the completeness proof is extended to this logic (sections 3, 7, 9, 15).

(3) The rules for term calculus are simplified (section 7).

(4) The completeness proof has been rewritten in order to make it more conspicuous. -

For useful discussions I am indebted to Dr. Rödding (1st edition) and to Dr. Ebbinghaus (2nd edition).

2. First order predicate logic with choice operator.

(For abbreviation we shall speak of "predicate logic".)

We use the following symbols:

(a) A denumerable set of r-place predicate symbols for each $r \geqslant 0$.

(b) A denumerable set of r-place function symbols for each $r \geqslant 0$.

(c) The logical symbols $\neg$ (not), $\wedge$ (and), $=$ (equals), ε (the selected example of).

(d) Parentheses.

The 0-place predicate symbols are called also proposition symbols, the 0-place function symbols subject symbols or variables. We use "x", "y", "z", "u", "v" as variables for variables, f as a variable for function symbols, P as a variable for predicate symbols.

Terms and formulas are introduced simultaneously by the following recursive definition:

(2.1) Every variable is a term.

(2.2) If f is an r-place function symbol ($r \geq 1$) and if $t_1, \ldots, t_r$ are terms, then $ft_1 \ldots t_r$ is a term.

(2.3) If α is a formula and x a variable, then $\varepsilon x \alpha$ is a term.

(2.4) If P is an r-place predicate symbol ($r \geq 0$) and $t_1, \ldots, t_r$ are terms, then $Pt_1 \ldots t_r$ is a formula.

(2.5) If t_1, t_2 are terms, then $t_1 = t_2$ is a formula.

(2.6) If α is a formula, then $\neg \alpha$ is a formula.

(2.7) If α and β are formulas, then $(\alpha \wedge \beta)$ is a formula.

We use "t", "s" as variables for terms and "α", "β", ... as variables for formulas.

A semantic basis for predicate logic is a quadruple

$$\mathfrak{B} = \langle \omega, \mathfrak{A}, \mathfrak{F}, a \rangle,$$

where

(2.8) ω is a non-empty individual domain. (We use "$\mathfrak{x}$" or "$\mathfrak{y}$" as a variable for the elements of ω.)

(2.9) $\mathfrak{A}$ is a set of attributes over ω. For each r there is at least one r-place attribute in $\mathfrak{A}$. If $\mathfrak{P}$ is an r-place attribute in $\mathfrak{A}$ and $\mathfrak{r}_1, \ldots, \mathfrak{r}_r$ are elements of ω, then it is defined whether $\mathfrak{A}$ holds for $\mathfrak{r}_1, \ldots, \mathfrak{r}_r$ (short $\mathfrak{P}\mathfrak{r}_1 \ldots \mathfrak{r}_r$) or not.

(2.10) $\mathcal{F}$ is a set of functions over ω. For each $r \geqslant 0$ there is at least one r-place function in $\mathcal{F}$. If φ is an r-place function in $\mathcal{F}$ and $\mathfrak{r}_1, \ldots, \mathfrak{r}_r$ are elements of ω, then $\varphi(\mathfrak{r}_1, \ldots, \mathfrak{r}_r)$ is an element of ω. The 0-place functions in $\mathcal{F}$ coincide with the elements of ω.

(2.11) $\mathfrak{a}$ is a choice operator on ω. If ρ is a subset of ω, then $\mathfrak{a}(\rho)$ is an element of ω. Moreover, if ρ is non-void, then $\mathfrak{a}(\rho)$ is an element of ρ.

An interpretation over $\mathfrak{B} = \langle \omega, \mathfrak{A}, \mathcal{F}, \mathfrak{a} \rangle$ is a mapping $\mathfrak{I}$ which maps every predicate symbol on an element of $\mathfrak{A}$ of the same place number and every function symbol on an element of $\mathcal{F}$ of the same place number.

We want to extend the domain of any given interpretation $\mathfrak{I}$ to the set of all terms and at the same time to introduce the relation Mod $\mathfrak{I}\alpha$ ($\mathfrak{I}$ is a model of α, or α holds for $\mathfrak{I}$). This is done by the following simultaneous recursion:

(2.1') $\mathfrak{I}(x)$ is already defined.

(2.2') $\mathfrak{I}(ft_1 \ldots t_r) = \mathfrak{I}(f)(\mathfrak{I}(t_1), \ldots, \mathfrak{I}(t_r))$.

(2.3') $\mathfrak{I}(\varepsilon x \alpha) = \mathfrak{a}(\{\mathfrak{r} : \text{Mod } \mathfrak{I}^{\mathfrak{r}}_x \alpha\})$, where $\mathfrak{I}^{\mathfrak{r}}_x$ is the interpretation over $\mathfrak{B}$ which has the same values as $\mathfrak{I}$ except for the argument x, which by $\mathfrak{I}^{\mathfrak{r}}_x$ is mapped on $\mathfrak{r}$. ($\{\mathfrak{r} : \ldots\}$ is the set of all $\mathfrak{r}$, for which ... .)

(2,4') Mod $\mathfrak{I}Pt_1 \ldots t_r$ iff $\mathfrak{I}(P)\mathfrak{I}(t_1) \ldots \mathfrak{I}(t_r)$ (i.e. $\mathfrak{I}(P)$ holds for $\mathfrak{I}(t_1) \ldots \mathfrak{I}(t_r)$).

(2.5') Mod $\mathfrak{I}\, t_1 = t_2$ iff $\mathfrak{I}(t_1) = \mathfrak{I}(t_2)$.

(2.6') Mod $\mathfrak{I} \neg \alpha$ iff not Mod $\mathfrak{I}\alpha$.

(2.7') Mod $\mathfrak{I}(\alpha \wedge \beta)$ iff Mod $\mathfrak{I}\alpha$ and Mod $\mathfrak{I}\beta$.

Let be $\mathfrak{M}$ a set of formulas. $\mathfrak{I}$ is a model of $\mathfrak{M}$, in symbols: Mod $\mathfrak{I}\mathfrak{M}$, if Mod $\mathfrak{I}\alpha$ for every $\alpha \in \mathfrak{M}$. α is a consequence of $\mathfrak{M}$, in symbols: $\mathfrak{M} \models \alpha$, if for every semantic basis $\mathfrak{B}$ and every interpretation $\mathfrak{I}$ over $\mathfrak{B}$ we have Mod $\mathfrak{I}\alpha$ if Mod $\mathfrak{I}\mathfrak{M}$.

3. First order term logic with choice operator.

(For abbreviation we shall speak of "term logic".)

In the fundamental definitions in the preceding section we have a kind of duplicity between terms and formulas. To avoid this duplicity we can try to use only one kind of expressions where the fundamental construction procedures (2.2), ..., (2.7) are reflected. We may distinguish between two different kinds of procedures. To the first class belongs procedure (2.3), where two new symbols ε and x are introduced. To the second class belong the procedures (2.2), (2.4), ..., (2.7), where only one symbol is introduced, namely f in (2.2), P in (2.4), $=$ in (2.5), $\neg$ in (2.6) and (neglecting parentheses) $\wedge$ in (2.7). To put the similarity of the procedures of the second class in greater evidence we may write "$= t_1 t_2$" in place of "$t_1 = t_2$" and "$\wedge \alpha \beta$" in place of "$(\alpha \wedge \beta)$". This is a hint that we may be able to collapse the procedures of the second class into one single procedure.

We want to construct a set of expressions using the procedure of the first class and the collapsed procedure of the second class. These expressions are called terms of the new language (and may not be confounded with the terms of the language which we have introduced in the preceding section).

We use the following symbols:

(a) A denumerable set of r-place function symbols for each $r \geqslant 0$. The 0-place function symbols are called variables. $\neg$ is a 1-place function symbol, $\wedge$ and $=$ are different 2-place function symbols.

(b) The choice symbol ε.

We use "x", "y", "z", "u", "v" as variables for variables and "f" as a variable for function symbols.

The terms are introduced by the following recursive definition:

(3.1) Every variable is a term.

(3.2) If f is an r-place function symbol ($r \geqslant 1$) and if $t_1, \ldots, t_r$ are terms, then $ft_1 \ldots t_r$ is a term.

(3.3) If t is a term and x a variable then $\varepsilon x t$ is a term.

We use "t", "s" as variables for terms (and also "α", "β", ..., starting with section 6).

A semantic basis for term logic is a septuple

$$\mathfrak{B} = \langle \omega, \pi, \mathcal{F}, \mathfrak{i}, \mathfrak{n}, \mathfrak{k}, \mathfrak{a} \rangle,$$

where

(3.4) ω is an individual domain with at least two elements.

(3.5) π is a subset of ω with $\pi \neq 0$ and $\omega - \pi \neq 0$.

(3.6) $\mathcal{F}$ is a set of functions over ω. For each $r \geqslant 0$ there is at least one r-place function in $\mathcal{F}$. If φ is an r-place function in $\mathcal{F}$ and $\mathfrak{x}_1, \ldots, \mathfrak{x}_r$ are elements of ω then $\varphi(\mathfrak{x}_1, \ldots, \mathfrak{x}_r)$ is an element of ω. The 0-place functions in $\mathcal{F}$ coincide with the elements of ω.

(3.7) $\mathfrak{i}$ is a 2-place function in $\mathcal{F}$. For each $\mathfrak{x}, \mathfrak{y} \in \omega$ we have

(*) $\qquad \mathfrak{i}(\mathfrak{x}, \mathfrak{y}) \in \pi \quad$ iff $\quad \mathfrak{x} = \mathfrak{y}$.

(3.8) $\mathfrak{n}$ is a 1-place function in $\mathcal{F}$. For each $\mathfrak{x} \in \omega$ we have

(**) $\qquad \mathfrak{n}(\mathfrak{x}) \in \pi \quad$ iff $\quad \mathfrak{x} \notin \pi$.

(3.9) $\mathfrak{k}$ is a 2-place function in $\mathcal{F}$. For each $\mathfrak{x}, \mathfrak{y} \in \omega$ we have

(***) $\qquad \mathfrak{k}(\mathfrak{x}, \mathfrak{y}) \in \pi \quad$ iff $\quad \mathfrak{x} \in \pi$ and $\mathfrak{y} \in \pi$.

(3.10) $\mathfrak{a}$ is a choice operator on ω. (As in 2.11.)

A semantic basis $\mathfrak{B}$ for term logic is called a restricted semantic basis, if π has exactly one element.

(*), (**), (***) reflect the essentials properties of identity, negation, conjunction, if the elements of π are considered as representing "the truth" and the elements of $\omega - \pi$ as representing "the falsehood".

An interpretation over $\mathfrak{B} = \langle \omega, \pi, \mathcal{F}, \mathfrak{i}, \mathfrak{n}, \mathfrak{k}, \mathfrak{a} \rangle$ is a mapping $\mathfrak{J}$ which maps every function symbol on an element of $\mathcal{F}$ of the same place number. In addition it is required that $\mathfrak{J}(=) = \mathfrak{i}$, $\mathfrak{J}(\neg) = \mathfrak{n}$, $\mathfrak{J}(\wedge) = \mathfrak{k}$. (Hence $=, \neg, \wedge$ may be considered as function constants.)

We extend the domain of any given interpretation $\mathfrak{J}$ to the set of all terms by the following recursion:

(3.1') $\mathfrak{I}(x)$ is already defined.

(3.2') $\mathfrak{I}(ft_1 \ldots t_r) = \mathfrak{I}(f)(\mathfrak{I}(t_1), \ldots, \mathfrak{I}(t_r))$.

(3.3') $\mathfrak{I}(\varepsilon xt) = \alpha(\{\mathfrak{r}: \mathfrak{I}_x^{\mathfrak{r}}(t) \in \pi\})$. ($\mathfrak{I}_x^{\mathfrak{r}}$ is defined in 2.3'.)

Let be $\mathfrak{M}$ a set of terms and t a term. t is a consequence of $\mathfrak{M}$, in symbols: $\mathfrak{M} \models t$, if for every semantic basis $\mathfrak{B}$ and every interpretation $\mathfrak{I}$ over $\mathfrak{B}$ we have $\mathfrak{I}(t) \in \pi$ if $\mathfrak{I}(\mathfrak{M}) \subset \pi$. ($\mathfrak{I}(\mathfrak{M})$ is the set of all $\mathfrak{I}(s)$ where $s \in \mathfrak{M}$.)

If we restrict ourselves to restricted semantic bases, where π has only one element, we get the notion of 1-consequence: t is a 1-consequence of $\mathfrak{M}$, in symbols: $\mathfrak{M} \models_1 t$, if for every restricted semantic basis $\mathfrak{B}$ and every interpretation $\mathfrak{I}$ over $\mathfrak{B}$ we have $\mathfrak{I}(t) \in \pi$ if $\mathfrak{I}(\mathfrak{M}) \subset \pi$. We have trivially

Theorem 3.1: If $\mathfrak{M} \models t$ then $\mathfrak{M} \models_1 t$.

The converse does not hold: If x, y are different then $\{x, y\} \models_1 =xy$ but not $\{x, y\} \models =xy$.

4. Embedding of predicate logic in term logic.

To each function symbol f of predicate logic we associate a function symbol $\bar{f}$ of term logic and to each predicate symbol P we associate a function symbol $\bar{P}$ of term logic. f and $\bar{f}$ have the same place number, also P and $\bar{P}$. In addition we require that the mapping $^{-}$ is an injection. Finally we assume that $=, \neg, \wedge$ do not belong to the range of $^{-}$.

We extend $^{-}$ to arbitrary terms (of predicate logic) and formulas as arguments by the following recursion (where $\equiv$ indicates the identity of expressions):

(1) $\bar{x}$ is already defined

(2) $\overline{ft_1 \ldots t_r} \equiv \bar{f}\bar{t}_1 \ldots \bar{t}_r$

(3) $\overline{\varepsilon x \alpha} \equiv \varepsilon \bar{x} \bar{\alpha}$

(4) $\overline{Pt_1 \ldots t_r} \equiv \bar{P}\bar{t}_1 \ldots \bar{t}_r$

(5) $\overline{t_1 = t_2} \equiv = \bar{t}_1 \bar{t}_2$

(6) $\overline{\neg \alpha} \equiv \neg \bar{\alpha}$

(7) $\overline{(\alpha \wedge \beta)} \equiv \wedge \bar{\alpha} \bar{\beta}$,

$\bar{t}$ and $\bar{\alpha}$ are terms of term logic. $\bar{\mathfrak{M}}$ is always the set of all $\bar{\alpha}$ with $\alpha \in \mathfrak{M}$.

Let now be

$$\mathfrak{B} = \langle \omega, \mathfrak{A}, \mathcal{F}, \mathfrak{a} \rangle$$

a semantic basis of predicate logic and

$$\bar{\mathfrak{B}} = \langle \omega, \pi, \bar{\mathcal{F}}, \mathfrak{i}, \mathfrak{n}, \mathfrak{t}, \mathfrak{a} \rangle$$

a semantic basis of term logic. ($\mathfrak{B}$ and $\bar{\mathfrak{B}}$ have the same ω and the same a.) Let be $\mathfrak{J}$ an interpretation over $\mathfrak{B}$ and $\bar{\mathfrak{J}}$ an interpretation over $\bar{\mathfrak{B}}$. Then we have

Lemma 4.1:

Let be

(i) $\bar{\mathfrak{J}}(\bar{f}) = \mathfrak{J}(f)$ for all function symbols of predicate logic, and

(ii) $\bar{\mathfrak{J}}(\bar{P})(\mathfrak{r}_1, \ldots, \mathfrak{r}_r) \in \pi$ iff $\mathfrak{J}(P)\mathfrak{r}_1 \ldots \mathfrak{r}_r$ for every predicate symbol and all $\mathfrak{r}_1, \ldots, \mathfrak{r}_r$.

Then

$\mathfrak{J}(t) = \bar{\mathfrak{J}}(\bar{t})$ for every term t of predicate logic and

Mod $\mathfrak{J}\alpha$ iff $\bar{\mathfrak{J}}(\bar{\alpha}) \in \pi$ for every formula α.

Proof: We show that the lemma holds for all $\mathfrak{J}, \bar{\mathfrak{J}}$ by induction on (2.1), ..., (2.7).

(1') $\quad \mathfrak{J}(x) = \bar{\mathfrak{J}}(\bar{x}) \quad$ by (i)

(2') $\quad \mathfrak{J}(ft_1 \ldots t_r) = \mathfrak{J}(f)(\mathfrak{J}(t_1), \ldots, \mathfrak{J}(t_r)) \quad$ by (2.2')

$= \bar{\mathfrak{J}}(\bar{f})(\bar{\mathfrak{J}}(\bar{t}_1), \ldots, \bar{\mathfrak{J}}(\bar{t}_r)) \quad$ by (i) and ind. hyp.

$= \bar{\mathfrak{J}}(\bar{f}\bar{t}_1 \ldots \bar{t}_r) \quad$ by (3.2')

$= \bar{\mathfrak{J}}(\overline{ft_1 \ldots t_r}) \quad$ by (2)

(3') $\quad \mathfrak{J}(\varepsilon x \alpha) = a(\{\mathfrak{r} : \text{Mod } \mathfrak{J}_x^{\mathfrak{r}} \alpha\}) \quad$ by (2.3')

$= a(\{\mathfrak{r} : \bar{\mathfrak{J}}_{\bar{x}}^{\mathfrak{r}}(\bar{\alpha}) \in \pi\}) \quad$ (see below)

$= \bar{\mathfrak{J}}(\varepsilon \bar{x} \bar{\alpha}) \quad$ (by (3.3')

$= \bar{\mathfrak{J}}(\overline{\varepsilon x \alpha}) \quad$ by (3)

In order to justify the transition from the first to the second line in (3') it is sufficient to show that for $\mathfrak{J}_x^{\mathfrak{r}}$ (in place of $\mathfrak{J}$) and $\bar{\mathfrak{J}}_{\bar{x}}^{\mathfrak{r}}$ (in place of $\bar{\mathfrak{J}}$) the assumptions (i) and (ii) of the lemma hold. (This enables us to apply the induction hypothesis.) It is only necessary to verify (i) for $f \equiv x$: $\bar{\mathfrak{J}}_{\bar{x}}^{\mathfrak{r}}(\bar{x}) = \mathfrak{r} = \mathfrak{J}_x^{\mathfrak{r}}(x)$.

(4')	<u>Mod</u> $\mathfrak{J}\ Pt_1 \ldots t_r$	iff	$\mathfrak{J}(P)\mathfrak{J}(t_1)\ldots\mathfrak{J}(t_r)$	by (2.4')
		iff	$\mathfrak{J}(P)\bar{\mathfrak{J}}(\bar{t}_1)\ldots\bar{\mathfrak{J}}(\bar{t}_r)$	by ind. hyp.
		iff	$\bar{\mathfrak{J}}(\bar{P})(\bar{\mathfrak{J}}(\bar{t}_1), \ldots, \bar{\mathfrak{J}}(\bar{t}_r)) \in \pi$	by (ii)
		iff	$\bar{\mathfrak{J}}(\bar{P}\bar{t}_1 \ldots \bar{t}_r) \in \pi$	by (3.2')
		iff	$\bar{\mathfrak{J}}(\overline{Pt_1 \ldots t_r}) \in \pi$	by (4)
(5')	<u>Mod</u> $\mathfrak{J}\ t_1 = t_2$	iff	$\mathfrak{J}(t_1) = \mathfrak{J}(t_2)$	by (2.5')
		iff	$\bar{\mathfrak{J}}(\bar{t}_1) = \bar{\mathfrak{J}}(\bar{t}_2)$	by ind. hyp.
		iff	$\mathfrak{i}(\bar{\mathfrak{J}}(\bar{t}_1), \bar{\mathfrak{J}}(\bar{t}_2)) \in \pi$	by (3.7) (*)
		iff	$\bar{\mathfrak{J}}(=)(\bar{\mathfrak{J}}(\bar{t}_1), \bar{\mathfrak{J}}(\bar{t}_2)) \in \pi$	$\bar{\mathfrak{J}}$ is an interpretation
		iff	$\bar{\mathfrak{J}}(=\bar{t}_1\bar{t}_2) \in \pi$	by (3.2')
		iff	$\bar{\mathfrak{J}}(\overline{t_1 = t_2}) \in \pi$	by (5)
(6')	<u>Mod</u> $\mathfrak{J}\ \neg\alpha$	iff	not <u>Mod</u> $\mathfrak{J}\ \alpha$	by (2.6')
		iff	$\bar{\mathfrak{J}}(\bar{\alpha}) \notin \pi$	by ind. hyp.
		iff	$\mathfrak{n}(\bar{\mathfrak{J}}(\bar{\alpha})) \in \pi$	by (3.8) (**)
		iff	$\bar{\mathfrak{J}}(\neg)\bar{\mathfrak{J}}(\bar{\alpha})) \in \pi$	$\bar{\mathfrak{J}}$ is an interpretation
		iff	$\bar{\mathfrak{J}}(\neg\bar{\alpha}) \in \pi$	by (3.2')
		iff	$\bar{\mathfrak{J}}(\overline{\neg\alpha}) \in \pi$	by (6)
(7')	<u>Mod</u> $\mathfrak{J}\ (\alpha \wedge \beta)$	iff	<u>Mod</u> $\mathfrak{J}\ \alpha$ and <u>Mod</u> $\mathfrak{J}\ \beta$	by (2.7')
		iff	$\bar{\mathfrak{J}}(\bar{\alpha}) \in \pi$ and $\bar{\mathfrak{J}}(\bar{\beta}) \in \pi$	by ind. hyp.
		iff	$\mathfrak{k}(\bar{\mathfrak{J}}(\bar{\alpha}), \bar{\mathfrak{J}}(\bar{\beta})) \in \pi$	by (3.9) (***)
		iff	$\bar{\mathfrak{J}}(\wedge)(\bar{\mathfrak{J}}(\bar{\alpha}), \bar{\mathfrak{J}}(\bar{\beta})) \in \pi$	$\bar{\mathfrak{J}}$ is an interpretation

iff $\bar{\mathfrak{J}}(\wedge \bar{\alpha}\,\bar{\beta}) \in \pi$ by (3.2')

iff $\bar{\mathfrak{J}}(\overline{\alpha \wedge \beta}) \in \pi$ by (7)

This lemma is used to prove the two following theorems.

Theorem 4.1: Let be α a formula and $\mathfrak{M}$ a set of formulas. Then we have

$$\text{If } \mathfrak{M} \models \alpha \quad \text{then} \quad \bar{\mathfrak{M}} \models \bar{\alpha}.$$

Proof: Let be $\bar{\mathfrak{J}}$ an interpretation over an arbitrary semantic basis $\bar{\mathfrak{B}} = \langle \omega, \pi, \bar{\mathcal{F}}, \mathfrak{i}, \mathfrak{n}, \mathfrak{k}, \mathfrak{a} \rangle$ with $\bar{\mathfrak{J}}(\bar{\mathfrak{M}}) \subset \pi$. We have to show that $\bar{\mathfrak{J}}(\bar{\alpha}) \in \pi$. For each P let be $\mathfrak{A}_P$ the attribute characterized by

$$\mathfrak{A}_P \mathfrak{r}_1 \ldots \mathfrak{r}_r \quad \text{iff} \quad \bar{\mathfrak{J}}(\bar{P})(\mathfrak{r}_1, \ldots, \mathfrak{r}_r) \in \pi$$

for all $\mathfrak{r}_1, \ldots, \mathfrak{r}_r \in \omega$. Now we choose $\mathfrak{B} = \langle \omega, \mathfrak{A}, \mathcal{F}, \mathfrak{a} \rangle$ such that every $\mathfrak{A}_P$ belongs to $\mathfrak{A}$ and every $\bar{\mathfrak{J}}(\bar{f})$ belongs to $\mathcal{F}$. Let the interpretation $\mathfrak{J}$ over $\mathfrak{B}$ be defined by stipulating $\mathfrak{J}(P) = \mathfrak{A}_P$ and $\mathfrak{J}(f) = \bar{\mathfrak{J}}(f)$. Then the assumptions of Lemma 4.1 are valid. Hence we have Mod $\mathfrak{J}\,\mathfrak{M}$ (since $\bar{\mathfrak{J}}(\bar{\mathfrak{M}}) \subset \pi$), Mod $\mathfrak{J}\,\alpha$ (since $\mathfrak{M} \models \alpha$) and therefore (using again the lemma) $\bar{\mathfrak{J}}(\bar{\alpha}) \in \pi$.

Theorem 4.2: Let be α a formula and $\mathfrak{M}$ a set of formulas. Then we have:

$$\text{If } \bar{\mathfrak{M}} \models_1 \bar{\alpha}, \quad \text{then} \quad \mathfrak{M} \cup \{\bigvee x \bigvee y \neg\, x = y\} \models \alpha .$$

Proof: Let be $\mathfrak{J}$ an interpretation over an arbitrary semantic basis $\mathfrak{B} = \langle \omega, \mathfrak{A}, \mathcal{F}, \mathfrak{a} \rangle$ with Mod $\mathfrak{J}\,\mathfrak{M} \cup \{\bigvee x \bigvee y \neg\, x = y\}$. (The existential quantifier $\bigvee$ is defined using ε as in section 1.) We have to show that Mod $\mathfrak{J}\,\alpha$.
Mod $\mathfrak{J}\,\bigvee x \bigvee y \neg x = y$ shows that ω has at least two elements. Let be $\mathfrak{r}_o$ an element of ω and $\pi = \{\mathfrak{r}_o\}$. Obviously there are functions $\mathfrak{i}, \mathfrak{n}, \mathfrak{k}$ (not necessarily in $\mathcal{F}$) with the properties:

$$\mathfrak{i}(\mathfrak{r}, \mathfrak{y}) = \mathfrak{r}_o \quad \text{iff} \quad \mathfrak{r} = \mathfrak{y} \qquad (\text{for every } \mathfrak{r}, \mathfrak{y} \in \omega)$$

$$\mathfrak{n}(\mathfrak{r}) = \mathfrak{r}_o \quad \text{iff} \quad \mathfrak{r} \neq \mathfrak{y}_o \qquad (\text{for every } \mathfrak{r} \in \omega)$$

$$\mathfrak{k}(\mathfrak{r}, \mathfrak{y}) = \mathfrak{r}_o \quad \text{iff} \quad \mathfrak{r} = \mathfrak{y} = \mathfrak{r}_o \qquad (\text{for every } \mathfrak{r}, \mathfrak{y} \in \omega).$$

For each P let be φ_P a function such that

$$\varphi_P(\mathfrak{r}_1, \ldots, \mathfrak{r}_r) = \mathfrak{r}_o \qquad \text{iff} \qquad \mathfrak{J}(P)\mathfrak{r}_1 \ldots \mathfrak{r}_r .$$

Let be $\bar{\mathcal{F}}$ a set of functions such that

every element of ω belongs to

$\mathfrak{i}, \mathfrak{n}, \mathfrak{k}$ belong to $\bar{\mathcal{F}}$,

$\mathfrak{J}(f) \in \bar{\mathcal{F}}$ for every f,

$\varphi_P \in \bar{\mathcal{F}}$ for every P.

Let be $\bar{\mathfrak{B}} = \langle \omega, \pi, \bar{\mathcal{F}}, \mathfrak{i}, \mathfrak{n}, \mathfrak{k}, \mathfrak{a} \rangle$. $\bar{\mathfrak{B}}$ is a restricted semantic basis. Let be $\bar{\mathfrak{J}}$ an interpretation over $\bar{\mathfrak{B}}$, where $\bar{\mathfrak{J}}(\bar{f}) = \mathfrak{J}(f)$ for every f and $\bar{\mathfrak{J}}(\bar{P}) = \varphi_P$ for every P. (The existence of $\bar{\mathfrak{J}}$ is guaranteed by the properties of the mapping $^{-}$.) Now Lemma 4.1 may be applied. We have $\bar{\mathfrak{J}}(\bar{\mathfrak{M}}) \subset \pi$ (since <u>Mod</u> $\mathfrak{J}$ $\mathfrak{M}$), $\bar{\mathfrak{J}}(\bar{\alpha}) \in \pi$ (since $\bar{\mathfrak{M}} \models_1 \bar{\alpha}$), hence <u>Mod</u> $\mathfrak{J}$ α.

From Theorem 3.1, 4.1 and 4.2 we conclude

<u>Theorem 4.3</u> : <u>Let be</u> $\mathfrak{M}$ <u>a set of formulas such that</u> $\mathfrak{M} \models \forall x \forall y \neg x = y$. <u>Then for each formula</u> α:

$$\mathfrak{M} \models \alpha \quad \underline{\text{iff}} \quad \bar{\mathfrak{M}} \models \bar{\alpha} \quad \underline{\text{and}} \quad \mathfrak{M} \models \alpha \quad \underline{\text{iff}} \quad \bar{\mathfrak{M}} \models_1 \bar{\alpha} .$$

<u>This shows that predicate logic is embeddable in term logic and in restricted term logic.</u>

5. Embedding of term logic in predicate logic.

To each function symbol f of term logic we associate a function symbol $\bar{f}$ of predicate logic such that f and $\bar{f}$ have the same place number. We require that $\bar{\ }$ is an injection. We abbreviate

$\bar{\equiv}$ by i, $\bar{\neg}$ by n, $\bar{\wedge}$ by k .

Let be P a fixed 1-place predicate symbol.

We extend the mapping $\bar{\ }$ to arbitrary terms of term logic by the following recursive definition:

(1) $\bar{x}$ is already defined

(2) $\overline{ft_1 \ldots t_r} \equiv \bar{f}\bar{t}_1 \ldots \bar{t}_r$

(3) $\overline{\varepsilon x t} \equiv \varepsilon \bar{x} P \bar{t}$.

$\bar{t}$ is a term of predicate logic. For each term t of term logic let be t_P the formula $P\bar{t}$. If $\mathfrak{M}$ is a set of terms of term logic, let be $\mathfrak{M}_P$ the set of all formulas t_P where $t \in \mathfrak{M}$.

Let now be

$$\mathfrak{B} = \langle \omega, \pi, \mathcal{F}, \mathfrak{i}, \mathfrak{n}, \mathfrak{k}, \mathfrak{a} \rangle$$

a semantic basis of term logic and

$$\bar{\mathfrak{B}} = \langle \omega, \mathfrak{A}, \bar{\mathcal{F}}, \mathfrak{a} \rangle$$

a semantic basis of predicate logic. ($\mathfrak{B}$ and $\bar{\mathfrak{B}}$ have the same ω and the same $\mathfrak{a}$.) Let be $\mathfrak{J}$ an interpretation over $\mathfrak{B}$ and $\bar{\mathfrak{J}}$ an interpretation over $\bar{\mathfrak{B}}$. Then we have

Lemma 5.1:

Let be

(i) $\bar{\mathfrak{J}}(\bar{f}) = \mathfrak{J}(f)$ <u>for every function symbol</u> f <u>of term logic, and</u>

(ii) $\bar{\mathfrak{J}}(P)\mathfrak{r}$ <u>iff</u> $\mathfrak{r} \in \pi$ <u>for every</u> $\mathfrak{r} \in \omega$.

Then

$$\mathfrak{J}(t) = \bar{\mathfrak{J}}(\bar{t}) \qquad \text{for every term } t \text{ of term logic.}$$

<u>Corollary</u> : <u>For every term</u> t <u>of term logic, under</u> (i), (ii) <u>we have</u>

$\mathfrak{J}(t) \in \pi$ <u>iff</u> <u>Mod</u> $\bar{\mathfrak{J}}\ t_P$,

for $\mathfrak{J}(t) \in \pi$	iff	$\bar{\mathfrak{J}}(P)\mathfrak{J}(t)$	by (ii)
	iff	$\bar{\mathfrak{J}}(P)\bar{\mathfrak{J}}(\bar{t})$	by Lemma 5.1
	iff	<u>Mod</u> $\bar{\mathfrak{J}}\ P\bar{t}$	by (2.4')
	iff	<u>Mod</u> $\bar{\mathfrak{J}}\ t_P$	by the definition of t_P.

<u>Proof of Lemma 5.1.</u> We show that the lemma holds <u>for all</u> $\mathfrak{J}$, $\bar{\mathfrak{J}}$ by induction on (3.1), (3.2), (3.3).

(1')	$\mathfrak{J}(x)$	$= \bar{\mathfrak{J}}(\bar{x})$	by (i)
(2')	$\mathfrak{J}(ft_1 \ldots t_r)$	$= \mathfrak{J}(f)\mathfrak{J}(t_1), \ldots, \mathfrak{J}(t_r))$	by (3.2')
		$= \bar{\mathfrak{J}}(\bar{f})\bar{\mathfrak{J}}(\bar{t}_1), \ldots, \bar{\mathfrak{J}}(\bar{t}_r))$	by (i) and ind. hyp.
		$= \bar{\mathfrak{J}}(\bar{f}\bar{t}_1 \ldots \bar{t}_r)$	by (2.2')
		$= \bar{\mathfrak{J}}(\overline{ft_1 \ldots t_r})$	by (2)
(3')	$\mathfrak{J}(\varepsilon xt)$	$= \alpha(\{\mathfrak{r} : \mathfrak{J}^{\mathfrak{r}}_{x}(t) \in \pi\})$	by (3.3')
		$= \alpha(\{\mathfrak{r} : \bar{\mathfrak{J}}^{\mathfrak{r}}_{\bar{x}}(\bar{t}) \in \pi\})$	(see below)
		$= \alpha(\{\mathfrak{r} : \bar{\mathfrak{J}}(P)\bar{\mathfrak{J}}^{\mathfrak{r}}_{\bar{x}}(\bar{t})\})$	by (ii)
		$= \alpha(\{\mathfrak{r} : \bar{\mathfrak{J}}^{\mathfrak{r}}_{\bar{x}}(P)\bar{\mathfrak{J}}^{\mathfrak{r}}_{\bar{x}}(\bar{t})\})$	since $\bar{\mathfrak{J}}(P) = \bar{\mathfrak{J}}^{\mathfrak{r}}_{\bar{x}}(P)$
		$= \alpha(\{\mathfrak{r} :$ <u>Mod</u> $\bar{\mathfrak{J}}^{\mathfrak{r}}_{\bar{x}} P\bar{t}\})$	by (2.4')

$= \bar{\mathfrak{J}}(\varepsilon\bar{x}P\bar{t})$ by (2.3')

$= \bar{\mathfrak{J}}(\overline{\varepsilon xt})$ by (3)

In order to justify the transition to the second line in (3') it is sufficient to verify that for $\mathfrak{J}_x^{\mathfrak{r}}$ (in place of $\mathfrak{J}$) and $\bar{\mathfrak{J}}_{\bar{x}}^{\mathfrak{r}}$ (in place of $\bar{\mathfrak{J}}$) the assumptions (i) and (ii) of the lemma hold. (This enables us to apply the induction hypothesis.)

Theorem 5.1 : Let be t a term and $\mathfrak{M}$ a set of terms of term logic. Then we have

If $\mathfrak{M}_P \models t_P$, then $\mathfrak{M} \models t$.

Proof: Let be $\mathfrak{J}$ an interpretation over an arbitrary semantic basis $\mathfrak{B} = \langle \omega, \pi, \mathfrak{A}, \mathcal{F}, \mathfrak{i}, \mathfrak{n}, \mathfrak{k}, \mathfrak{a} \rangle$ with $\mathfrak{J}(\mathfrak{M}) \subset \pi$. We have to show that $\mathfrak{J}(t) \in \pi$. There is a semantic basis $\bar{\mathfrak{B}} = \langle \omega, \mathfrak{A}, \bar{\mathcal{F}}, \mathfrak{a} \rangle$ such that (1) the 1-place attribute $\mathfrak{P}$ which holds exactly for the elements of π is an element of $\mathfrak{A}$ and (2) for every function symbol f of term logic the function $\mathfrak{J}(f)$ belongs to $\bar{\mathcal{F}}$. Let be $\bar{\mathfrak{J}}$ an interpretation over $\bar{\mathfrak{B}}$ with $\bar{\mathfrak{J}}(P) = \mathfrak{P}$ and $\bar{\mathfrak{J}}(\bar{f}) = \mathfrak{J}(f)$ for every function symbol f of term logic. We now apply the Corollary and have Mod $\bar{\mathfrak{J}}\ \mathfrak{M}_P$ (since $\mathfrak{J}(\mathfrak{M}) \subset \pi$), Mod $\bar{\mathfrak{J}}\ t_P$ (since $\mathfrak{M}_P \models t_P$) and finally (using again the Corollary) $\mathfrak{J}(t) \in \pi$.

Theorem 5.2 : Let be t a term and $\mathfrak{M}$ a set of terms of term logic. Let be

$\mathfrak{M}'_P = \mathfrak{M}_P \cup \{\bigvee xPx, \bigvee x\neg Px, \bigwedge x(Pnx \leftrightarrow \neg Px), \bigwedge x\bigwedge y(Pkxy \leftrightarrow Px \wedge Py),$

$\bigwedge x\bigwedge y(Pixy \leftrightarrow x = y)\}$. Then we have

If $\mathfrak{M} \models t$ then $\mathfrak{M}'_P \models t_P$.

Proof: Let be $\bar{\mathfrak{J}}$ an interpretation over an arbitrary semantic basis $\bar{\mathfrak{B}} = \langle \omega, \mathfrak{A}, \bar{\mathcal{F}}, \mathfrak{a} \rangle$ with Mod $\bar{\mathfrak{J}}\ \mathfrak{M}'_P$. We have to show that Mod $\bar{\mathfrak{J}}\ t_P$.

Let be π the set of all $\mathfrak{r} \in \omega$ with $\bar{\mathfrak{J}}(P)\mathfrak{r}$. π and $\omega - \pi$ are not void since Mod $\bar{\mathfrak{J}}\{\bigvee xPx, \bigvee x\neg Px\}$.

Let $\mathcal{F}$ be a set of functions such that

every element of ω belongs to $\mathcal{F}$,

$\bar{\mathfrak{J}}(\bar{f}) \in \mathcal{F}$ for each function symbol f of term logic.

Let be

$$\mathfrak{i} = \bar{\mathfrak{J}}(i) \qquad (= \bar{\mathfrak{J}}(=)),$$
$$\mathfrak{n} = \bar{\mathfrak{J}}(n) \qquad (= \bar{\mathfrak{J}}(\neg)),$$
$$\mathfrak{k} = \bar{\mathfrak{J}}(k) \qquad (= \bar{\mathfrak{J}}(\wedge)).$$

Then $\mathfrak{i}, \mathfrak{n}, \mathfrak{k} \in \mathcal{F}$. Conditions (*), (**), (***) of section 3 hold since Mod $\bar{\mathfrak{J}}\ \mathfrak{M}'_P$.

$\mathfrak{B} = \langle \omega, \pi, \mathcal{F}, \mathfrak{i}, \mathfrak{n}, \mathfrak{k}, \alpha \rangle$ is a semantic basis. Let be $\mathfrak{J}$ an interpretation over $\mathfrak{B}$ such that $\mathfrak{J}(f) = \bar{\mathfrak{J}}(\bar{f})$. (As always required for an interpretation in term logic, we have $\mathfrak{J}(\neg) = \mathfrak{n}$ etc.) We now apply the Corollary and have $\mathfrak{J}(\mathfrak{M}) \subset \pi$ (since Mod $\bar{\mathfrak{J}}\ \mathfrak{M}_P$), $\mathfrak{J}(t) \in \pi$ (since $\mathfrak{M} \models t$, and finally (using again the Corollary) Mod $\bar{\mathfrak{J}}\ t_P$.

The same proof gives

Theorem 5.2' : Let be t a term and $\mathfrak{M}$ a set of terms of term logic. Let be

$\mathfrak{M}''_P = \mathfrak{M}_P \cup \{\bigvee x Px, \bigvee x \neg Px, \bigwedge x(Pnx \leftrightarrow \neg Px), \bigwedge x \bigwedge y(Pkxy \leftrightarrow Px \wedge Py),$

$\bigwedge x \bigwedge y(Pixy \leftrightarrow x = y), \bigwedge x \bigwedge y(Px \wedge Py \rightarrow x = y)$. Then we have

If $\mathfrak{M} \models_1 t$, then $\mathfrak{M}''_P \models t_P$.

As immediate consequences we get the two following theorems.

Theorem 5.3 : Let be $\mathfrak{M}$ a set of terms of term logic. Let

$\bigvee x Px$, $\bigvee x \neg Px$, $\bigwedge x(Pnx \leftrightarrow \neg Px)$, $\bigwedge x \bigwedge y(Pkxy \leftrightarrow Px \wedge Py)$, $\bigwedge x \bigwedge y(Pixy \leftrightarrow x = y)$

be consequences of $\mathfrak{M}_P$. Then we have for each term t of term logic

$\mathfrak{M} \models t$ iff $\mathfrak{M}_P \models t_P$.

Theorem 5.3' : Let be $\mathfrak{M}$ a set of terms of term logic. Let be the formulas listed in Theorem 5.3 consequences of $\mathfrak{M}_P$, and in addition the formula $\bigwedge x \bigwedge y(Px \wedge Py \rightarrow x = y)$. Then we have for each term t of term logic

$\mathfrak{M} \models_1 t$ iff $\mathfrak{M}_P \models t_P$.

These theorems show that term logic and restricted term logic are embeddable in predicate logic.

Together with the final result of the last section we get a theorem concerning predicate logic, which shows that (under certain assumptions) it is possible to replace an axiom system by an equivalent set of axioms where the propositional connectives occur in a very simple fashion.

6. Free occurence of a variable. Rank of a term. Elementary substitution.

Starting with this section we deal only with term logic (and not with predicate logic). For convenience we use not only "s", "t", but also "α", "β", ... as variables for terms.

In preparation to the formulation of the rules of term calculus in section 7 we introduce some useful concepts.

<u>Free</u> x α means that the variable x occurs free in the term α . This relation is introduced by recursion:

(F1) <u>Free</u> x u iff x ≡ u

(F2) <u>Free</u> x $ft_1 \ldots t_r$ iff <u>Free</u> x t_j for at least one j

(F3) <u>Free</u> x εuα iff <u>Free</u> x and x ≢ u.

The <u>rank</u> R(α) of a term is the number of (ε-symbols or function symbols with place number > 0) occuring in α:

(R1) $R(u) = 0$

(R2) $R(ft_1 \ldots t_r) = 1 + R(t_1) + \ldots + R(t_r) \qquad (r \geq 1)$

(R3) $R(\varepsilon u \alpha) = 1 + R(\alpha)$.

<u>Esubst</u> α x t β means that from the term α we obtain the term β by the elementary substitution of the term t for the variable x. This relation is defined by recursion on R(α):

(E1) <u>Esubst</u> u x t β

<u>iff</u> (u ≡ x and t ≡ β) or (u ≢ x and u ≡ β)

(E2) <u>Esubst</u> $f\alpha_1 \ldots \alpha_r$ x t β

<u>iff</u> there exist $\beta_1, \ldots, \beta_r$ such that <u>Esubst</u> α_j x t β_j $(j = 1, \ldots, r)$ and $\beta \equiv f\beta_1 \ldots \beta_r$.

(E3) $\underline{\text{Esubst}}$ $\varepsilon u\alpha$ x t β

$\underline{\text{iff}}$ (not $\underline{\text{Free}}$ x $\varepsilon u\alpha$ and $\beta \equiv \varepsilon u\alpha$)

or ($\underline{\text{Free}}$ x $\varepsilon u\alpha$ and not $\underline{\text{Free}}$ u t and there exists a term β'

such that $\underline{\text{Esubst}}$ α x t β' and $\beta \equiv \varepsilon u\beta'$)

Not for each α, x, t there exists a term β such that $\underline{\text{Esubst}}$ α x t β, since in (E3) the case $\underline{\text{Free}}$ x $\varepsilon u\alpha$ and $\underline{\text{Free}}$ u t does not occur. This is remedied in the (generalized) substitution of section 11.

Now we enumerate four theorems. These are proved in analogy to the usual predicate logic (cf. e.g. [4] (Bibliography)). Only for (6.3) we indicate some details of the proof.

(6.0) (<u>Theorem of coincidence</u>) If $\mathfrak{J}_1$ and $\mathfrak{J}_2$ are interpretations over the same semantical basis which have the same values (i.e. coincide) for every variable which occurs free in the term α and for every function symbol of place number $r \geq 1$ which occurs in α, then $\mathfrak{J}_1(\alpha) = \mathfrak{J}_2(\alpha)$.

(6.1) (<u>Theorem of invariance</u>) If not $\underline{\text{Free}}$ x α, then $\underline{\text{Esubst}}$ α x t α.

(6.2) (<u>Theorem concerning the freedom of variables</u>) Let be $\underline{\text{Esubst}}$ α x t β. Then we have for every variable u:

$\underline{\text{Free}}$ u β iff ($\underline{\text{Free}}$ u α and $u \not\equiv x$)

or ($\underline{\text{Free}}$ x α and $\underline{\text{Free}}$ u t)

(6.3) (<u>Theorem of transition</u>) Let be $\underline{\text{Esubst}}$ α x t β. Then

$$\mathfrak{J}_x^{\mathfrak{J}(t)}(\alpha) = \mathfrak{J}(\beta) .$$

That this holds <u>for every interpretation</u> $\mathfrak{J}$ can be proved by recursion on the structure of α. We indicate the proof for the case where we have $\underline{\text{Free}}$ x $\varepsilon u\alpha'$, not $\underline{\text{Free}}$ u t, $\underline{\text{Esubst}}$ α' x t β', $\underline{\text{Esubst}}$ $\varepsilon u\alpha'$ x t $\varepsilon u\beta'$. We have to show that $\mathfrak{J}_x^{\mathfrak{J}(t)}(\varepsilon u\alpha') = \mathfrak{J}(\varepsilon u\beta')$, i.e. that

$$\mathfrak{a}(\{\mathfrak{r} : \mathfrak{J}^{\mathfrak{J}(t)\mathfrak{r}}_{x\ \ u}(\alpha')\} \in \pi) = \mathfrak{a}(\{\mathfrak{r} : \mathfrak{J}^{\mathfrak{r}}_{u}(\beta') \in \pi\}) .$$

It is sufficient to show that $\mathfrak{J}^{\mathfrak{J}(t)\mathfrak{r}}_{x\ \ u}(\alpha') = \mathfrak{J}^{\mathfrak{r}}_{u}(\beta')$ for every $\mathfrak{r}$. By induction hypothesis, applied on the interpretation $\mathfrak{J}^{\mathfrak{r}}_{u}$, we get $\mathfrak{J}^{\mathfrak{r}\,\mathfrak{J}(t)}_{ux} = \mathfrak{J}^{\mathfrak{r}}_{u}(\beta')$. It remains to check that

$$\mathfrak{J}^{\mathfrak{J}(t)\mathfrak{r}}_{x\ \ u} = \mathfrak{J}^{\mathfrak{r}\,\mathfrak{J}(t)}_{ux} .$$

This is obvious if $x \not\equiv u$. But $x \equiv u$ is impossible, since <u>Free</u> x $\varepsilon u \alpha'$ but not <u>Free</u> u $\varepsilon u \alpha'$.

7. A term calculus.

A sequent is void or an expression of the form $\alpha_1 \ldots \alpha_r$ $(r \geq 1)$ where every α_j is a term. We use "Σ" as a variable for sequences.

If $\alpha_1 \ldots \alpha_r \equiv \beta_1 \ldots \beta_s$ then $r = s$ and $\alpha_j \equiv \beta_j$ for every j $(1 \leq j \leq r)$. The α_j are called the members of the sequent $\alpha_1 \ldots \alpha_r$. The void sequent has no member.

The rules of the calculus enable us to derive sequents. We first enumerate the rules. Explanations will be given afterwards.

(K_o) Introduction of the conjunction

$$\frac{\begin{array}{l}\Sigma_1 \ \alpha \\ \Sigma_2 \ \beta\end{array}}{\Sigma_{12} \ \wedge\alpha\beta}$$

(K_1) Elimination of the conjunction (first rule)

$$\frac{\Sigma \ \wedge\alpha\beta}{\Sigma \ \alpha}$$

(K_2) Elimination of the conjunction (second rule)

$$\frac{\Sigma \ \wedge\alpha\beta}{\Sigma \ \beta}$$

(E) Exhaustion

$$\frac{\begin{array}{l}\Sigma_1\alpha \ \beta \\ \Sigma_2\neg\alpha \ \beta\end{array}}{\Sigma_{12} \ \beta}$$

(C) Contradiction

$$\frac{\Sigma_1 \quad \alpha \qquad \Sigma_2 \quad \neg\alpha}{\Sigma_{12} \quad \beta}$$

(I_x) Identity (first rule)

$$\frac{}{=xx}$$

(I_x^t) Identity (second rule)

$$\frac{\Sigma \quad \alpha}{\Sigma = xt \quad \beta}, \quad \text{if } \underline{\text{Esubst}}\ \alpha\ x\ t\ \beta$$

(S_x^t) Elementary substitution

$$\frac{\Sigma \quad \alpha}{\Sigma' \quad \alpha'}, \quad \text{if } \underline{\text{Esubst}}\ \Sigma\alpha\ x\ t\ \Sigma'\alpha'$$

(B_x^y) Renaming of bound variables

$$\frac{}{= \varepsilon x \alpha \varepsilon y \beta}, \quad \text{if } \underline{\text{Esubst}}\ \alpha\ x\ y\ \beta \text{ and not } \underline{\text{Free}}\ y\ \alpha$$

(P_x) Particularization

$$\frac{}{\alpha \quad \beta}, \quad \text{if } \underline{\text{Esubst}}\ \alpha\ x\ \varepsilon x\alpha\ \beta$$

(E_x) Extensionality

$$\frac{\Sigma_1\alpha \quad \beta \qquad \Sigma_2\beta \quad \alpha}{\Sigma_{12} \quad = \varepsilon x \alpha \varepsilon x \beta}, \quad \text{if not } \underline{\text{Free}}\ x\ \Sigma_{12}$$

These rules constitute the term calculus. The restricted term calculus has in addition the following rule:

(R_x^y) Restriction (only for restricted term calculus)

$$\overline{\quad xy \quad = xy \quad}$$

Remarks: The rules (I_x), (B_x^y), (P_x) and (R_x^y) permit to write down the sequents below the line (since nothing is presupposed above the line). The other rules permit the transition from one sequent or two sequents (premises) which are indicated above the line, to the new sequent indicated below the line. In rule (K_o) the first premise is a sequent of the form $\Sigma_1\alpha$ and the second premise a sequent of the form $\Sigma_2\beta$ (where Σ_1 or Σ_2 may be void). (K_o) permits the transition from these two sequents to a new sequent. The last member of the new sequent is $\wedge\alpha\beta$. The preceding part Σ_{12} of the new sequent is any sequent which has the following property: An arbitrary term γ is a member of Σ_{12} iff γ is a member of Σ_1 or (vel) a member of Σ_2. (If Σ_1 and Σ_2 are not void there are infinitely many sequents $\Sigma_{12}\wedge\alpha\beta$ which can be obtained by an application of (K_o) to the sequents $\Sigma_1\alpha$ and $\Sigma_2\beta$.) What has been said about (K_o) may be stated in a similar way for (E), (C) and (E_x).

In (S_x^t) the condition Esubst $\Sigma\alpha$ x t $\Sigma'\alpha'$ means there are terms $\gamma_1, \ldots, \gamma_r$ and $\gamma'_1, \ldots, \gamma'_r$ such that $\Sigma \equiv \gamma_1 \ldots \gamma_r$ and $\Sigma' \equiv \gamma'_1 \ldots \gamma'_r$ and that Esubst γ_j x t γ'_j $(j = 1, \ldots, r)$ and Esubst α x t α'.

A sequent Σ is called derivable (resp. 1-derivable), in symbols $\vdash\Sigma$ resp. $\vdash_1\Sigma$, if Σ can be obtained by a finite number of applications of the rules without resp. with rule (B_x^y). A term α is called derivable (1-derivable) from a set $\mathfrak{M}$ of terms, in symbols $\mathfrak{M}\vdash\alpha$ $(\mathfrak{M}\vdash_1\alpha)$, if there is a sequent Σ whose members are elements of $\mathfrak{M}$, such that $\vdash\Sigma\alpha$ $(\vdash_1\Sigma\alpha)$.

8. Soundness of the rules.

Let be Σ a sequent and $\mathfrak{M}$ the set of its members. A sequent $\Sigma\alpha$ is called sound (1-sound) if $\mathfrak{M} \models \alpha$ ($\mathfrak{M} \models_1 \alpha$). A sound sequent is 1-sound.

A rule is sound (1-sound), if applied to sound (1-sound) premises, it leads to a sound (1-sound) sequent. (A rule without premises is sound (1-sound) if every sequent obtainable by it is a sound (1-sound) sequent.)

Theorem 8.1: Every rule, except (R_x^y), is sound.
Every rule (included (R_x^y)) is 1-sound.

This theorem gives immediately

Theorem 8.2: (Soundness of term calculus)

If $\mathfrak{M} \vdash \alpha$ then $\mathfrak{M} \models \alpha$. If $\mathfrak{M} \vdash_1 \alpha$ then $\mathfrak{M} \models_1 \alpha$.

Proof of the Theorem 8.1. We show only that (K_1), (I_x^t), (B_x^y), (P_x), (E_x) are sound and that (R_x^y) is 1-sound but not sound.

Soundness of (K_1) : Let be $\mathfrak{J}(\Sigma) \subset \pi$ (i.e. $\mathfrak{J}(\gamma) \in \pi$ for each member γ of Σ). Since by assumption $\Sigma \wedge \alpha\beta$ is sound, we have $\mathfrak{J}(\wedge\alpha\beta) \in \pi$, whence $\mathfrak{t}(\mathfrak{J}(\alpha), \mathfrak{J}(\beta)) \in \pi$ and $\mathfrak{J}(\alpha) \in \pi$.

Soundness of (I_x^t) : Let be $\mathfrak{J}(\Sigma) \subset \pi$ and $\mathfrak{J}(=xt) \in \pi$. We get $\mathfrak{J}(\alpha) \in \pi$ using the soundness of $\Sigma\alpha$. From $\mathfrak{J}(=xt) \in \pi$ we obtain $\mathfrak{J}(x) = \mathfrak{J}(t)$. We now have $\mathfrak{J}(\beta) = \mathfrak{J}_x^{\mathfrak{J}(t)}(\alpha)$ (by the theorem of transition (6.3)) $= \mathfrak{J}_x^{\mathfrak{J}(x)}(\alpha) = \mathfrak{J}(\alpha)$. Hence $\mathfrak{J}(\beta) \in \pi$.

Soundness of (B_x^y) : It is sufficient to show that

$$\{\mathfrak{r} : \mathfrak{J}_x^{\mathfrak{r}}(\alpha) \in \pi\} = \{\mathfrak{r} : \mathfrak{J}_y^{\mathfrak{r}}(\beta) \in \pi\}.$$

This holds if $\mathfrak{J}_x^{\mathfrak{r}}(\alpha) = \mathfrak{J}_y^{\mathfrak{r}}(\beta)$ for all $\mathfrak{r}$. We have $\mathfrak{J}_y^{\mathfrak{r}}(\beta) = \mathfrak{J}_y^{\mathfrak{r}}{}^{\mathfrak{J}_y^{\mathfrak{r}}(y)}(\alpha)$ (by (6.3)) =

$\mathfrak{I}^{\mathfrak{r}\,\mathfrak{r}}_{y\,x}(\alpha) = \mathfrak{I}^{\mathfrak{r}}_{x}(\alpha)$ (by (6.0), since not Free y α).

Soundness of (P_x) : (This rule correspondends to the introduction of the existential quantifier in the usual predicate logic, cf. section 1.) Let be $\mathfrak{I}(\alpha) \in \pi$. From (6.3) we get $\mathfrak{I}(\beta) = \mathfrak{I}^{\mathfrak{I}(\varepsilon x\alpha)}_{x}(\alpha)$. Furthermore we have $\mathfrak{I}(\varepsilon x\alpha) = \mathfrak{a}(\{\mathfrak{r} : \mathfrak{I}^{\mathfrak{r}}_{x}(\alpha) \in \pi\})$. Here $\mathfrak{a}$ is applied on a non-void class ρ, since $\mathfrak{I}(x) \in \rho$ (because $\mathfrak{I}^{\mathfrak{I}(x)}_{x}(\alpha) = \mathfrak{I}(\alpha) \in \pi$). Hence $\mathfrak{I}(\varepsilon x\alpha)$ is an element of ρ, i.e. $\mathfrak{I}^{\mathfrak{I}(\varepsilon x\alpha)}_{x}(\alpha) \in \pi$.

Soundness of (E_x): Let be $\mathfrak{I}(\Sigma_{12}) \subset \pi$. We have to show that $\mathfrak{I}(=\varepsilon x\alpha\varepsilon x\beta) \in \pi$, i.e. that $\mathfrak{I}(\varepsilon x\alpha) = \mathfrak{I}(\varepsilon x\beta)$. For this it is sufficient to prove that for each $\mathfrak{r}$: $\mathfrak{I}^{\mathfrak{r}}_{x}(\alpha) \in \pi$ iff $\mathfrak{I}^{\mathfrak{r}}_{x}(\beta) \in \pi$. Let be $\mathfrak{I}^{\mathfrak{r}}_{x}(\alpha) \in \pi$. We have $\mathfrak{I}(\Sigma_1) \subset \pi$ and by (6.0) also $\mathfrak{I}^{\mathfrak{r}}_{x}(\Sigma_1) \subset \pi$ (since not Free x Σ_1). Now the soundness of $\Sigma_1\alpha\beta$ gives $\mathfrak{I}^{\mathfrak{r}}_{x}(\beta) \in \pi$. The converse statement is shown in the same way.

1-Soundness of (R^y_x) : Let be $\mathfrak{I}(x) \in \pi$ and $\mathfrak{I}(y) \in \pi$. π has only one element in a restricted semantic basis. Hence $\mathfrak{I}(x) = \mathfrak{I}(y)$ and $\mathfrak{I}(=xy) \in \pi$.

(R^y_x) is not sound. This is evident by taking a semantic basis where π has two different elements $\mathfrak{r}, \mathfrak{y}$ and using an interpretation $\mathfrak{I}$ over this basis with $\mathfrak{I}(x) = \mathfrak{r}$ and $\mathfrak{I}(y) = \mathfrak{y}$.

9. Survey of the completeness proof.

A. The following sections deal with the proof for

Theorem 9.1: (Completeness of term calculus and restricted term calculus)

If $\mathfrak{M} \models \alpha$, *then* $\mathfrak{M} \vdash \alpha$. *If* $\mathfrak{M} \models_1 \alpha$, *then* $\mathfrak{M} \vdash_1 \alpha$.

This together with Theorem 8.2 leads to

Theorem 9.2: (Equivalence of $\models$ and $\vdash$ and of $\models_1$ and $\vdash_1$)

$\mathfrak{M} \models \alpha$ *iff* $\mathfrak{M} \vdash \alpha$. $\mathfrak{M} \models_1 \alpha$ iff $\mathfrak{M} \vdash_1 \alpha$.

Theorem 9.1 will be proved primary for $\models$ and $\vdash$ (in sections 9 - 14) using the wellknown method of Henkin and Hasenjaeger. For $\models_1$ and $\vdash_1$ see section 15.

B. Satisfiability and Consistency.

Definition 9.1 : An interpretation $\mathfrak{J}$ satisfies a set $\mathfrak{M}$ of terms if $\mathfrak{J}(\mathfrak{M}) \subset \pi$. $\mathfrak{M}$ is satisfiable if there is an interpretation $\mathfrak{J}$ which satisfies $\mathfrak{M}$.

Definition 9.2 : A set $\mathfrak{M}$ of terms is consistent if there exists a term which is not derivable from $\mathfrak{M}$. $\mathfrak{M}$ is inconsistent if $\mathfrak{M}$ is not consistent.

C. Following Gödel the contraposition of the proposition of Theorem 9.1 is proved in three steps:

(a) not $\mathfrak{M} \vdash \alpha$ (assumption)

(b) $\mathfrak{M} \cup \{\neg\alpha\}$ is consistent

(c) $\mathfrak{M} \cup \{\neg\alpha\}$ is satisfiable

(d) not $\mathfrak{M} \models \alpha$.

The transition from (c) to (d) is obvious. That (b) is a consequence of (a) is shown by using the propositional part of the rules of term calculus (see (12.4)). The essential part of the proof is the transition from (b) to (c). It can be shown that every consistent set $\mathfrak{M}$ is satisfiable. The proof of this Theorem of satisfiability is prepared in sections 10, 11, 12, outlined in section 13 and completed in section 14.

In order to prove that every consistent set $\mathfrak{M}$ is satisfiable we shall assume in addition that there are infinitely many variables which are not free in any element of $\mathfrak{M}$. In D we shall show that this is not a essential restriction.

D. Isomorphisms. Let the function φ map the set of variables one-one into itself. The image of x by φ is the variable x^{φ}. If in a given term t each variable x which occurs in t (free or not) is replaced by x^{φ} we obtain again a term which is designated by t^{φ}. If $\mathfrak{M}$ is a set of terms, $\mathfrak{M}^{\varphi}$ is the set of t^{φ} where $t \in \mathfrak{M}$. We have

(9.1) Free $x\,\alpha$ iff Free $x^{\varphi}\,\alpha^{\varphi}$.

(9.2) If Esubst $\alpha\, x\, t\, \beta$, then Esubst $\alpha^{\varphi}\, x^{\varphi}\, t^{\varphi}\, \beta^{\varphi}$.

(9.3) If $\mathfrak{M} \vdash \alpha$, then $\mathfrak{M}^{\varphi} \vdash \alpha^{\varphi}$.

(9.1) and (9.2) are obvious. For (9.3), check the rules in order to show that from a given derivation we obtain again a derivation if we replace everything by its φ-image.

(9.4) If $\mathfrak{M}^{\varphi} \vdash \alpha^{\varphi}$ then $\mathfrak{M} \vdash \alpha$.

Proof: Since $\mathfrak{M}^{\varphi} \vdash \alpha^{\varphi}$ there are elements $\alpha_1, \ldots, \alpha_r \in \mathfrak{M}$ such that $\vdash \alpha_1^{\varphi} \ldots \alpha_r^{\varphi} \alpha^{\varphi}$. There is only a finite number of variables occuring in the sequent $\alpha_1 \ldots \alpha_r \alpha$. Hence there is a one-one mapping ψ of the set of all variables into itself such that $x^{\varphi\psi} \equiv x$ for each variable x which occurs in $\alpha_1 \ldots \alpha_r \alpha$. From $\{\alpha_1^{\varphi}, \ldots, \alpha_r^{\varphi}\} \vdash \alpha^{\varphi}$ using (9.3) we get $\{\alpha_1^{\varphi\psi}, \ldots, \alpha_r^{\varphi\psi}\} \vdash \alpha^{\varphi\psi}$, i.e. $\{\alpha_1, \ldots, \alpha_r\} \vdash \alpha$. Hence $\mathfrak{M} \vdash \alpha$.

(9.5) If $\mathfrak{M}$ is consistent then $\mathfrak{M}^{\varphi}$ is consistent.

Otherwise $\mathfrak{M}^{\varphi} \vdash \alpha^{\varphi}$ for each α, and $\mathfrak{M} \vdash \alpha$ for each α (by (9.5)) which contradicts the consistency of $\mathfrak{M}$.

Let be $\mathfrak{J}$ an interpretation over a semantical basis $\mathfrak{B}$. Let be $\mathfrak{J}^{\varphi}$ the interpretation over $\mathfrak{B}$ where

$$\mathfrak{J}^{\varphi}(x) = \mathfrak{J}(x^{\varphi}) \qquad \text{for every } x,$$

$$\mathfrak{J}^{\varphi}(f) = \mathfrak{J}(f) \qquad \text{for each } r\text{-place } f \ (r \geq 1).$$

It is easy to show by induction on the rank of t that

(9.6) $$\mathfrak{J}^{\varphi}(t) = \mathfrak{J}(t^{\varphi})$$

for every interpretation $\mathfrak{J}$. We restrict ourselves to the case $t \equiv \varepsilon x \alpha$. Here we have

$$\mathfrak{J}^{\varphi}(\varepsilon x \alpha) = \mathfrak{a}(\{\mathfrak{r} : \mathfrak{J}^{\varphi \mathfrak{r}}_{x}(\alpha) \in \pi\}),$$

$$\mathfrak{J}((\varepsilon x \alpha)^{\varphi}) = \mathfrak{J}(\varepsilon x^{\varphi} \alpha^{\varphi}) = \mathfrak{a}(\{\mathfrak{r} : \mathfrak{J}^{\mathfrak{r}}_{x^{\varphi}}(\alpha^{\varphi}) \in \pi\}).$$

$\mathfrak{J}^{\mathfrak{r}}_{x^{\varphi}}(\alpha^{\varphi}) = (\mathfrak{J}^{\mathfrak{r}}_{x^{\varphi}})^{\varphi}(\alpha)$ by induction hypothesis. Hence it is sufficient to verify that

$(\mathfrak{J}^{\mathfrak{r}}_{x^{\varphi}})^{\varphi} = \mathfrak{J}^{\varphi \mathfrak{r}}_{x}$. -

From (9.6) we get immediately

(9.7) If $\mathfrak{M}^{\varphi}$ is satisfiable then $\mathfrak{M}$ is satisfiable.

Later we use a mapping φ where each variable (which is characterized by an index) is mapped on a variable with an even index. Hence in $\mathfrak{M}^{\varphi}$ we have no variable with an odd index. Now (9.5) and (9.7) make sure that in C it is allowed to assume that there are infinitely many variables which are not free in any element of $\mathfrak{M}$.

10. Some theorems concerning deducibility.

Only for some of these propositions we indicate a proof. (Cf. also Hermes [4].

(10. 1) $\vdash \alpha\alpha$.

Use (P_x) where not Free $x\ \alpha$. Cf. (6. 1).

(10. 2) If $\vdash \Sigma\alpha$ then $\vdash \Sigma'\alpha$, if each member of Σ is a member of Σ'.

Use (K_o), (K_1) and (10. 1).

(10. 3) If $\vdash \Sigma_1\alpha$ and $\vdash \Sigma_2\alpha\beta$ then $\vdash \Sigma_{12}\beta$ (syllogism).

Use (10. 1), (W), (E) .

(10. 4) $\vdash {=}tt$.

(10. 5) $\vdash {=}t_1t_2\ {=}t_2t_1$.

(10. 6) $\vdash {=}t_1t_2\ {=}t_2t_3\ {=}t_1t_3$.

(10. 7) $\vdash {=}s_1t_1\ \dots\ {=}s_rt_r\ \ {=}fs_1\dots s_rft_1\dots t_r$.

(10. 8) $\vdash {=}st\ s\ t$ and $\vdash {=}st\ t\ s$.

Use (10. 1), the second rule for identity, the rule of substitution and (10. 2).

(10. 9) If $\vdash {=}\alpha\beta$ then $\vdash {=}\varepsilon x\alpha\varepsilon x\beta$.

Use (10. 8) and (E_x).

11. Substitution.

A. Survey. As remarked in section 6 not for every α, x, t there is a β such that Esubst $\alpha\ x\ t\ \beta$. Here we want to associate with every α, x, t in an effective way a term $\alpha\frac{t}{x}$ such that (among other properties) $\alpha\frac{t}{x} \equiv \beta$, if Esubst $\alpha\ x\ t\ \beta$. Hence we may consider the operation $\frac{t}{x}$ of substitution as a generalization of the elementary substitution. The connection between elementary substitution and substitution may be described by the following statements: First we associate with every α, x, t in an effective way a term α_x^t (not $\alpha\frac{t}{x}$). α and α_x^t are equivalent (i.e. $\vdash = \alpha\alpha_x^t$) and we have Esubst $\alpha_x^t\ x\ t\ \alpha\frac{t}{x}$. If there is a β such that Esubst $\alpha\ x\ t\ \beta$, then $\alpha_x^t \equiv \alpha$ and $\alpha\frac{t}{x} \equiv \beta$. We obtain α_x^t from α by "renaming bound variables" (cf. B).

In section C some of the theorems of section 6 are extended to the substitution.

Finally in section D we want to show that from the rules of the calculus where elementary substitution is used (those are the rules (I_x^t), (S_x^t), (B_x^y) and (P_x)) we obtain correct rules if we replace elementary substitution by substitution.

B. Simultaneous definitions for α_x^t and $\alpha\frac{t}{x}$. These definitions proceed by induction on the rank of α. At the same time we show that always

(11.1) $\vdash = \alpha\alpha_x^t$,

(11.2) Esubst $\alpha_x^t\ x\ t\ \alpha\frac{t}{x}$,

(11.3) α and α_x^t have the same free variables,

(11.4) if t_1 and t_2 have the same free variables then

$$\alpha_x^{t_1} \equiv \alpha_x^{t_2} ,$$

(11.5) $R(\alpha\frac{y}{x}) = R(\alpha)$,

(11.6) if Esubst α x t β then $\alpha_x^{\,t} \equiv \alpha$ and $\alpha\frac{t}{x} \equiv \beta$.

B1'. $\alpha \equiv x$. We define $\alpha_x^{\,t} \equiv \alpha$ and $\alpha\frac{t}{x} \equiv t$. (11.1), ..., (11.6) are trivial.

B1''. $\alpha \equiv z$, $z \not\equiv x$. We define $\alpha_x^{\,t} \equiv \alpha\frac{t}{x} \equiv \alpha$. (11.1), ..., (11.6) are trivial.

B2. $\alpha \equiv f\alpha_1 \ldots \alpha_r$. We define $\alpha_x^{\,t} \equiv f\alpha_{1x}^{\;\,t} \ldots \alpha_{rx}^{\;\,t}$ and $\alpha\frac{t}{x} \equiv f\alpha_1\frac{t}{x} \ldots \alpha_r\frac{t}{x}$.

In order to show (11.1) use (10.7). (11.2), ..., (11.6) are trivial.

B3'. $\alpha \equiv \varepsilon u\beta$, not Free x α. We define $\alpha_x^{\,t} \equiv \alpha\frac{t}{x} \equiv \alpha$. (11.1), ..., (11.6) are trivial.

B3''. $\alpha \equiv \varepsilon u\beta$, Free x α, not Free u t. We define $\alpha_x^{\,t} \equiv \varepsilon u\beta_x^{\,t}$ and $\alpha\frac{t}{x} \equiv \varepsilon u\beta\frac{t}{x}$.

(11.1), ..., (11.6) are obtained easily using the induction hypothesis. For (11.1) use in addition (10.9).

B3'''. $\alpha \equiv \varepsilon u\beta$, Free x α, Free u t. Let be v the variable with minimal index where not Free v α and not Free v t. We define

$$[\varepsilon u\beta]_x^{\,t} \equiv \varepsilon v\beta\frac{v}{u}{}_x^{t} \qquad \text{and} \qquad [\varepsilon u\beta]\frac{t}{x} \equiv \varepsilon v\beta\frac{v}{u}\frac{t}{x}.$$

(Here $\beta\frac{v}{u}{}_x^{t}$ and $\beta\frac{v}{u}\frac{t}{x}$ may be considered as already defined since $R(\beta\frac{v}{u}) = R(\beta)$ by induction hypothesis (11.5).)

ad (11.1): Using (10.6) and (10.3) it is sufficient to show

(a) $\vdash\ = \varepsilon u \beta \varepsilon u \beta^{v}_{u}$,

(b) $\vdash\ = \varepsilon u \beta^{v}_{u} \varepsilon v \beta\frac{v}{u}$,

(c) $\vdash\ = \varepsilon v \beta\frac{v}{u}\, \varepsilon v \beta\frac{v\ t}{u\ x}$.

We get (a) by the induction hypothesis $\vdash\ = \beta\beta^{v}_{u}$ using (10.9).

We get (b) using (U^{v}_{u}) since by induction hypothesis (1) Esubst β^{v}_{u} u v $\beta\frac{v}{u}$ and (2) not Free v β^{v}_{u} (otherwise Free v β by (11.3), hence $v \equiv u$ since not Free v $\varepsilon u \beta$; but we have $v \not\equiv u$ since Free u t and not Free v t.

We get (c) by the induction hypothesis $\vdash\ = \beta\frac{v}{u}\beta\frac{v\ t}{u\ x}$ using (10.9).

ad (11.2): We get this by section 6, (E_3) from Free x $\varepsilon v \beta\frac{v\ t}{u\ x}$ (cf. later), not Free v t and the induction hypothesis Esubst $\beta\frac{v\ t}{u\ x}$ x t $\beta\frac{v\ t}{u\ x}$ (use $R(\beta\frac{v}{u}) = R(\beta)$). - In order to show Free x $\varepsilon v \beta\frac{v\ t}{u\ x}$ one may verify successively: Free x $\varepsilon u \beta$, Free x β and $x \not\equiv u$, Free x β^{v}_{u} (11.3), Free x $\beta\frac{v}{u}$ (from Free x α and not Free v α we get $x \not\equiv v$; now use (6.2)), Free x $\beta\frac{v\ t}{u\ x}$ (induction hypothesis), Free x $\varepsilon v \beta\frac{v\ t}{u\ x}$ (since $x \not\equiv v$).

ad (11.3): The following propositions are equivalent :

Free z $\varepsilon v \beta\frac{v\ t}{u\ x}$

$z \not\equiv v$ and Free z $\beta\frac{v\ t}{u\ x}$

$z \not\equiv v$ and Free z $\beta\frac{v}{u}$

$z \not\equiv v$ and Free z β^{v}_{u} and $z \not\equiv u$ (using (6.2))

$z \neq v$ and Free $z\ \beta$ and $z \neq u$

Free $z\ \beta$ and $z \neq u$ (from this we get $z \neq v$ since not Free $v\ \beta$, Free $z\ \varepsilon u\beta$).

ad (11.4): Not Free $v\ \alpha$ and not Free $v\ t_1$ is equivalent to not Free $v\ \alpha$ and not Free $v\ t_2$.

ad (11.5): $R(\varepsilon v\beta\frac{v}{u}\frac{y}{x}) = 1 + R(\beta\frac{v}{u}\frac{y}{x}) = 1 + R(\beta) = R(\varepsilon u\beta)$.

ad (11.6): Under the assumption B3''' there is no γ with Esubst $\alpha\ x\ t\ \gamma$.

C. Generalization of the theorems formulated in section 6 for elementary substitution.

(11.7) (Theorem of invariance) If not Free $x\ \alpha$ then $\alpha\frac{t}{x} \equiv \alpha$.

(11.8) (Theorem concerning the freedom of variables)

Free $u\ \alpha\frac{t}{x}$ iff (Free $u\ \alpha$ and $u \neq x$) or (Free $x\ \alpha$ and Free $u\ t$).

(11.9) Corollary $\alpha\frac{\varepsilon x\beta}{x}\frac{t}{x} \equiv \alpha\frac{\varepsilon x\beta}{x}$.

(11.10) (Theorem of transition)

$$\mathfrak{J}_x^{\mathfrak{J}(t)}(\alpha) = \mathfrak{J}(\alpha\frac{t}{x}).$$

We get (11.7) using (6.1) and (11.6). We get (11.8) using (11.2), (11.3), (6.2). From (11.8) we get not Free $x\ \alpha\frac{\varepsilon x\beta}{x}$. Hence we obtain (11.9) from (11.7).

We get (11.10) using (11.1), (11.2) and (6.3).

D. <u>The generalized rules</u> (i_x^t), (s_x^t), (b_x^y), (p_x).

(i_x^t) <u>Identity (generalized rule)</u>

$$\frac{\Sigma \quad \alpha}{\Sigma = xt \quad \alpha_x^t}$$

(s_x^t) <u>Substitution</u>

$$\frac{\Sigma \quad \alpha}{\Sigma_x^t \quad \alpha\frac{t}{x}}$$

(b_x^y) <u>Renaming of bound variables (generalized rule)</u>

$$\frac{}{=\varepsilon x\alpha\,\varepsilon y\alpha\frac{y}{x}}\quad,\ \text{if not } \underline{\text{Free}}\ y\ \alpha$$

(p_x) <u>Particularization (generalized rule)</u>

$$\frac{}{\alpha \quad \alpha\frac{\varepsilon x\alpha}{x}}$$

<u>Justification of</u> (i_x^t) by the following derivation (in a similar way we may justify (s_x^t)):

$\Sigma \quad \alpha$ (assumption)

$=\alpha\alpha_x^t$ (11.1)

$$\alpha\ \alpha_x^t \qquad (10.8), (10.3)$$

$$\Sigma\ \alpha_x^t \qquad (10.3)$$

$$\Sigma = xt\ \alpha\frac{t}{x} \qquad (I_x^t),\ (11.2)$$

<u>Justification of</u> b_x^y by the following derivation :

$$=\alpha\alpha_x^y \qquad (11.1)$$

$$=\varepsilon x\alpha\varepsilon x\alpha_x^y \qquad (10.9)$$

$$=\varepsilon x\alpha_x^y\varepsilon y\alpha\frac{y}{x} \qquad (B_x^y)$$

$$=\varepsilon x\alpha\varepsilon y\alpha\frac{y}{x} \qquad (10.6),\ (10.3)$$

<u>Justification of</u> (p_x) : We abbreviate $\alpha_x^{\varepsilon x\alpha}$ by β. α and β have the same free variables (11.3), hence also $\varepsilon x\alpha$ and $\varepsilon x\beta$. By (11.4) we get $\alpha_x^{\varepsilon x\alpha} \equiv \alpha_x^{\varepsilon x\beta}$, i.e. $\beta \equiv \alpha_x^{\varepsilon x\beta}$. By (11.2) <u>Esubst</u> $\alpha_x^{\varepsilon x\beta}$ x $\varepsilon x\beta\ \alpha\frac{\varepsilon x\beta}{x}$, hence <u>Esubst</u> β x $\varepsilon x\beta\ \alpha\frac{\varepsilon x\beta}{x}$. Now we give the following derivation:

$$\beta\ \alpha\frac{\varepsilon x\beta}{x} \qquad (P_x) \text{ with } \underline{\text{Esubst}}\ \beta\ x\ \varepsilon x\beta\ \alpha\frac{\varepsilon x\beta}{x}$$

$$=\alpha\beta \qquad (11.1)$$

$$=\beta\alpha \qquad (10.5), (10.3)$$

$=\varepsilon x \beta \varepsilon x \alpha$ (10. 9)

$\beta \quad \beta$ (10. 1)

$\beta \;=x\varepsilon x\alpha \;\; \alpha\frac{\varepsilon x\alpha}{x}$ ($I_x^{\varepsilon x\alpha}$) with <u>Esubst</u> $\beta \; x \; \varepsilon x\alpha \; \alpha\frac{\varepsilon x\alpha}{x}$

$\alpha\frac{\varepsilon x\beta}{x} \;=\varepsilon x\beta\varepsilon x\alpha \;\; \alpha\frac{\varepsilon x\alpha}{x}$ ($S_x^{\varepsilon x\beta}$) with <u>Esubst</u> $\beta \; x \; \varepsilon x\beta \; \alpha\frac{\varepsilon x\beta}{x}$

and (11. 9)

$\alpha\frac{\varepsilon x\beta}{x} \;\; \alpha\frac{\varepsilon x\alpha}{x}$ (10. 3)

$\beta \;\; \alpha\frac{\varepsilon x\alpha}{x}$ (10. 3)

$\alpha \quad \beta$ (10. 8), (10. 3)

$\alpha \;\; \alpha\frac{\varepsilon x\alpha}{x}$ (10. 3)

12. Further theorems concerning deducibility.

A. About propositional connectives.

(12.1) $\vdash \wedge\alpha\beta\ \alpha$ and $\vdash \wedge\alpha\beta\ \beta$.

Use (10.1), (K_1) resp. (K_2) .

(12.2) $\vdash \alpha\ \beta\ \wedge\alpha\beta$.

Use (10.1), (K_0).

(12.3) If $\mathfrak{M} \vdash \alpha$ and $\mathfrak{M} \vdash \neg\alpha$ then $\mathfrak{M}$ is inconsistent.

Use (W).

(12.4) If $\mathfrak{M} \cup \{\alpha\}$ is inconsistent then $\mathfrak{M} \vdash \neg\alpha$.

If $\mathfrak{M} \cup \{\neg\alpha\}$ is inconsistent then $\mathfrak{M} \vdash \alpha$.

Use (10.1) and (E) .

(12.5) If $\mathfrak{M} \cup \{\alpha\}$ and $\mathfrak{M} \cup \{\neg\alpha\}$ are inconsistent then $\mathfrak{M}$ is inconsistent .

Use (12.4), (12.3).

(12.6) If $\mathfrak{M} \cup \{\alpha\}$ is inconsistent and $\mathfrak{M} \vdash \alpha$ then $\mathfrak{M}$ is inconsistent.

Use (12.4), (12.3).

(12.7) If $\mathfrak{M} \cup \{=xs\}$ is inconsistent and x is not free in $\mathfrak{M} \cup \{s\}$ then $\mathfrak{M}$ is inconsistent.

Proof: Since $\mathfrak{M} \cup \{=xs\}$ is inconsistent there are sequents Σ_1, Σ_2 whose members are elements of $\mathfrak{M}$ such that $\vdash \Sigma_1 = xs\ s$ and $\vdash \Sigma_2 = xs \neg s$ (cf. (10.2)). Using (S_x^s) we get $\vdash \Sigma_1 = ss\ s$ (cf. (6.1)). Using (10.4) and (10.3) we have $\vdash \Sigma_1\ s$. In a similar way we obtain $\vdash \Sigma_2 \neg s$. Now (12.3) gives (12.7).

(12.8) If $\mathfrak{M}_o \subset \mathfrak{M}_1 \subset \mathfrak{M}_2 \subset \ldots$ and if every $\mathfrak{M}_j$ is consistent then $\mathfrak{M} = \bigcup_j \mathfrak{M}_j$ is consistent.

(If $\mathfrak{M} \vdash \alpha$ then $\mathfrak{E} \vdash \alpha$ for a finite subset $\mathfrak{E}$ of $\mathfrak{M}$.)

B. About substitution.

(12.9) $\vdash = xt\ = ss\frac{t}{x}$.

Let be $v \not\equiv x$, and not Free $v\ s$ and not Free $v\ t$. Then by (11.8) not Free $v\ s\frac{t}{x}$. We have the following derivation:

$= vs\ = vs$	(10.1)
$= vs\ = xt\ = vs\frac{t}{x}$	(i_x^t)
$= ss\ = xt\ = ss\frac{t}{x}$	(s_v^s), (11.7)
$= xt\ = ss\frac{t}{x}$	(10.3)

(12.10) $\vdash = uv\ = s\frac{u}{x}s\frac{v}{x}$, if $u \not\equiv x$, $v \not\equiv x$.

This is shown by the following derivation:

$= xu\ = ss\frac{u}{x}$	(12.9)
$= xv\ = ss\frac{v}{x}$	(12.9)
$= xu\ = uv\ = s\frac{u}{x}s\frac{v}{x}$	(10.5), (10.6), (10.3)

$$=uu \quad =uv \quad =s\frac{u}{x}s\frac{v}{x} \qquad (s_x^u),\ (11.8)$$

$$=uv \quad =s\frac{u}{x}s\frac{v}{x}$$

(12.11) $\qquad \vdash\ =s\frac{t}{x}s\frac{u}{x}\frac{t}{u}$, if $u \not\equiv x$ and not $\underline{\text{Free}}$ u s.

Let be $v \not\equiv x$, $v \not\equiv u$ and not $\underline{\text{Free}}$ v t. We have the following derivation:

$$=vu \quad =s\frac{v}{x}s\frac{u}{x} \qquad (12.10)$$

$$=vt \quad =s\frac{v}{x}s\frac{u}{x}\frac{t}{u} \qquad (s_u^t)$$

$$=tt \quad =s\frac{v}{x}\frac{t}{v}s\frac{u}{x}\frac{t}{u} \qquad (s_v^t)$$

$$=s\frac{v}{x}\frac{t}{v}s\frac{u}{x}\frac{t}{u} \qquad (10.3)$$

$$=xv \quad =ss\frac{v}{x} \qquad (12.9)$$

$$=tv \quad =s\frac{t}{x}s\frac{v}{x} \qquad (s_x^t)$$

$$=tt \quad =s\frac{t}{x}s\frac{v}{x}\frac{t}{v} \qquad (s_v^t)$$

$$=s\frac{t}{x}s\frac{v}{x}\frac{t}{v} \qquad (10.3)$$

$$=s\frac{t}{x}s\frac{u}{x}\frac{t}{u} \qquad (10.6),\ (10.3)$$

(12.12) $\qquad \vdash\ =t_1t_2 \quad =s\frac{t_1}{x}s\frac{t_2}{x}$.

Let be $u \not\equiv v$, $u \not\equiv x$, $v \not\equiv x$, not free u and v in s or t_1 or t_2. We have the following derivation:

$$=uv \quad =s\frac{u}{x}s\frac{v}{x} \qquad (12.10)$$

$$=t_1v \quad =s\frac{u}{x}\frac{t_1}{u}s\frac{v}{x} \qquad (s_u^{t_1})$$

$$=t_1t_2 \quad = s\frac{u}{x}\frac{t_1}{u}s\frac{v}{x}\frac{t_2}{v} \qquad (s\frac{t_2}{v})$$

$$=t_1t_2 \quad = s\frac{t_1}{x}s\frac{t_2}{x} \qquad (12.11), (10.6), (10.5), (10.3)$$

C. Using E_x .

(12.13), ..., (12.17) are preparations for (12.18).

(12.13) (Contraposition) If $\vdash \neg\alpha\ \neg\beta$ then $\vdash \beta\alpha$.

Use (10.1), (W), (E).

(12.14) (Contraposition) If $\vdash \Sigma\ \neg\alpha\ \beta$ then $\vdash \Sigma\ \neg\beta\ \alpha$.

Use (10.1), (W), (E).

(12.15) $\vdash \neg\wedge\alpha\neg\beta\ \alpha\ \beta$.

Use (12.2), (10.2), (12.14) .

(12.16) $\leftrightarrow\alpha\beta$ is an abbreviation for $\wedge\neg\wedge\alpha\neg\beta\neg\wedge\beta\neg\alpha$.

(12.17) $\vdash \leftrightarrow\alpha\beta\ \alpha\ \beta$ and $\vdash \leftrightarrow\alpha\beta\ \beta\ \alpha$.

Use (12.1), (12.15), (10.3).

(12.18) $\vdash [\leftrightarrow\alpha_1\alpha_2]\ \frac{\varepsilon x \nleftrightarrow \alpha_1\alpha_2}{x} = \varepsilon x\alpha_1\,\varepsilon x\alpha_2$

We have the following derivation

$$\neg\leftrightarrow\alpha_1\alpha_2 \quad [\neg\leftrightarrow\alpha_1\alpha_2]\frac{\varepsilon x\neg\leftrightarrow\alpha_1\alpha_2}{x} \qquad (p_x)$$

$$\neg\leftrightarrow\alpha_1\alpha_2 \quad \neg[\leftrightarrow\alpha_1\alpha_2]\frac{\varepsilon x\neg\leftrightarrow\alpha_1\alpha_2}{x} \qquad \text{(sect. 11, B2)}$$

$$[\leftrightarrow\alpha_1\alpha_2]\frac{\varepsilon x\neg\leftrightarrow\alpha_1\alpha_2}{x} \quad \leftrightarrow\alpha_1\alpha_2 \qquad (12.13)$$

$$[\leftrightarrow\alpha_1\alpha_2]\frac{\varepsilon x\neg\leftrightarrow\alpha_1\alpha_2}{x}\ \alpha_2 \quad \alpha_2 \qquad (12.17), (10.3)$$

$$[\leftrightarrow\alpha_1\alpha_2]\frac{\varepsilon x\neg\leftrightarrow\alpha_1\alpha_2}{x}\ \alpha_1 \quad \alpha_1 \qquad (12.17), (10.3)$$

$$[\leftrightarrow\alpha_1\alpha_2]\frac{\varepsilon x\neg\leftrightarrow\alpha_1\alpha_2}{x} \quad = \varepsilon x\alpha_1\,\varepsilon x\alpha_2 \qquad (E_x)$$

By (11.8) x is not free in $[\leftrightarrow\alpha_1\alpha_2]\frac{\varepsilon x\neg\leftrightarrow\alpha_1\alpha_2}{x}$. Hence the last transition is permitted.

13. Survey of the proof of the Theorem on satisfiability.

In section 9,C it was stated that we get the completeness of our calculus if we have the following

(13.1) Theorem on satisfiability: Let be $\mathfrak{M}$ a consistent set of terms. We assume that there exists an infinite number of variables not occuring free in any element of $\mathfrak{M}$. Then $\mathfrak{M}$ is satisfiable.

In this section we want to give a survey of the proof for (13.1). The details are treated in section 14.

First step: Maximalization of $\mathfrak{M}$. Let be $s_o, s_1, s_2, \ldots$ a denumeration of all terms. We define inductively a sequence $\mathfrak{M}_j$ of sets of terms as follows:

(13.2) $$\mathfrak{M}_o = \mathfrak{M}$$

(13.3) $$\mathfrak{M}_{j+1} = \begin{cases} \mathfrak{M}_j \cup \{=xs_j\}, & \text{if } \mathfrak{M} \cup \{s_j\} \text{ is inconsistent} \\ \mathfrak{M}_j \cup \{s_j\} \cup \{=xs_j\}, & \text{if } \mathfrak{M}_j \cup \{s_j\} \text{ is consistent}. \end{cases}$$

In order to determinate the variable x in (13.3) in an unique way we postulate that x is the variable with the least index which does not occur free in any element of $\mathfrak{M}_j \cup \{s_j\}$.

Let $\mathfrak{M}^*$ be the union of all $\mathfrak{M}_j$.

From (12.7) we induce that every $\mathfrak{M}_j$ is consistent. Now we get with (12.8):

(13.5) $\mathfrak{M}^*$ is consistent.

We want to give some other properties of $\mathfrak{M}^*$.

(13.6) If $\alpha \notin \mathfrak{M}^*$ then $\mathfrak{M}^* \cup \{\alpha\}$ is inconsistent.

(13.2), (13.3)

(13.7) If $\mathfrak{M}^* \vdash \alpha$ then $\alpha \in \mathfrak{M}^*$.

Otherwise by (13.6) $\mathfrak{M}^* \cup \{\alpha\}$ and by (12.6) $\mathfrak{M}^*$ itself would be inconsistent,which contradicts (13.5).

(13.8) If $\vdash \alpha_1 \dots \alpha_n \alpha$ and $\alpha_1, \dots, \alpha_n \in \mathfrak{M}^*$ then $\alpha \in \mathfrak{M}^*$.

We first get $\mathfrak{M}^* \vdash \alpha$ and then apply (13.7).

(13.8') If $\vdash =\alpha\beta$ then ($a \in \mathfrak{M}^*$ iff $\beta \in \mathfrak{M}^*$).

(10.8), (13.8)

(13.9) $\neg\alpha \in \mathfrak{M}^*$ iff $\alpha \notin \mathfrak{M}^*$.

If $\alpha \in \mathfrak{M}^*$ and $\neg\alpha \in \mathfrak{M}^*$ then by (12.3) $\mathfrak{M}^*$ would be inconsistent, contradicting (13.5). If $\alpha \notin \mathfrak{M}^*$ and $\neg\alpha \notin \mathfrak{M}^*$ then $\mathfrak{M}^*$ could be inconsistent by (13.6) and (12.5).

(13.10) $\wedge\alpha\beta \in \mathfrak{M}^*$ iff ($\alpha \in \mathfrak{M}^*$ and $\beta \in \mathfrak{M}^*$).

(12.1), (12.2), (13.8).

(13.11) For each term s there is a variable x such that $=xs \in \mathfrak{M}^*$.

By (13.3).

Second step: Definition of a semantic basis $\langle \omega, \pi, \mathfrak{F}, \mathfrak{l}, \mathfrak{n}, \mathfrak{t}, \mathfrak{a} \rangle$.

We introduce a binary relation $\sim$ between terms by the

(13.12) Definition: $s \sim t$ iff $=st \in M^*$.

Using (10.4), (10.5), (10.6), (13.8) we see that $\sim$ is an equivalence relation.

(13.13) Definition: $\bar{s}$ is the class of all terms which are equivalent to s.

(13.14) Each class $\bar{s}$ has at least one variable as an element.

By (13.11).

(13.15) Definition: ω is the set of all classes $\bar{s}$.

(13.16) Definition: π is the set of all classes $\bar{s}$ such that $s \in \mathfrak{M}^*$.

This definition requires to show that it is independent of the representative s of $\bar{s}$. Indeed we have: If $\bar{s} = \bar{t}$ and $s \in \mathfrak{M}^*$ then $=st \in \mathfrak{M}^*$, $\vdash =st\ s\ t$ (10.8), hence $t \in \mathfrak{M}^*$ by (13.8).

(13.17) π and ω-π are not void.

$\vdash =xx$, hence $=xx \in \mathfrak{M}^*$ (13.7) and $\overline{=xx} \in \pi$. $\overline{\neg =xx} \notin \pi$. Otherwise $\overline{\neg =xx} \in \pi$ (13.9), $\neg =xx \in \mathfrak{M}^*$, which together with $=xx \in \mathfrak{M}^*$ would contradict the consistency of $\mathfrak{M}^*$.

We now associate with every function symbol f a function $\tilde{f}$. f and $\tilde{f}$ have the same place number r. $\tilde{f}$ is defined for each r-tuple of classes. Its value is a class according to the

(13.18) Definition: $\tilde{f}(\bar{t}_1, \ldots, \bar{t}_r) = \overline{f(t_1, \ldots, t_r)}$.

In section 14, A we show that this definition is independent of the representatives.

(13.19) Definition: $\mathcal{F}$ is the set of all functions $\tilde{f}$. Furthermore

$$\mathfrak{i} = \tilde{=}, \quad \mathfrak{n} = \tilde{\neg}, \quad \mathfrak{t} = \tilde{\wedge}.$$

In section 14, B we show that (*), (**), (***) of section 3 hold for $\mathfrak{i}, \mathfrak{n}, \mathfrak{t}$. - The 0-place functions in $\mathcal{F}$ are by (13.18) the classes $\bar{x}$, hence by (13.14) exactly the elements of ω, as required in (3.6).

The most interesting point is the definition of the choice operator α. For this purpose we introduce the concept of a representable subclass ρ of ω. For the proof of (13.1) it is only relevant how $\alpha(\rho)$ is defined for a representable subclass ρ.

(13.20) A subclass ρ of ω is called representable iff there is a variable x and a term s such that

$$(*) \qquad \bar{t} \in \rho \quad \text{iff} \quad s\frac{t}{x} \in \mathfrak{M}^*$$

for every term t.

If we have (*) for every t, ρ is determined by x and s. We then call x, s a pair of representatives of ρ. There is at most a denumerable set of pairs of representatives for ρ. Hence if ω is infinite, not every subclass ρ of ω is a representable subclass of ω. The void subclass is representable. x, $\neg$=xx is a pair of representatives. If ρ is representable there are infinitely many pairs of representatives for ρ. -

(13.21) Definition: If ρ is representable and if x, s is a pair of representatives for ρ, then $\alpha(\rho) = \overline{\varepsilon xs}$.

(13.22) Definition: If ρ is not representable then (e.g) let be $\alpha(\rho)$ the class $\bar{s}$, where s is lexicographically the shortest term such that $\bar{s} \in \rho$.

In section 14, C we show that (13.21) is independent of the choice of the pair x, s of representatives. If $\rho \neq o$, then $\alpha(\rho) \in \rho$. This is evident for (13.22) and will be shown for (13.21) in section 14, D.

By the definitions (13.15), . . . , (13.22) we have given a semantical basis

$$\mathfrak{B} = \langle \omega, \pi, \mathfrak{F}, \mathfrak{i}, \mathfrak{n}, \mathfrak{k}, \mathfrak{a} \rangle .$$

Third step: Definition of an interpretation $\mathfrak{J}$ over $\mathfrak{B}$, which satisfies $\mathfrak{M}^*$ (and also $\mathfrak{M}$, since $\mathfrak{M} \subset \mathfrak{M}^*$).

(13.23) Definition: $\mathfrak{J}(f) = \tilde{f}$.

Then by (13.19) $\mathfrak{J}(=) = \mathfrak{i}$, $\mathfrak{J}(\neg) = \mathfrak{n}$, $\mathfrak{J}(\wedge) = \mathfrak{k}$. Hence $\mathfrak{J}$ is an interpretation over $\mathfrak{B}$. In section 14, D we show that

(13.24) $\mathfrak{J}(t) = \bar{t}$ for every term t.

If now $t \in \mathfrak{M}^*$ then $\bar{t} \in \pi$ (13.16) and $\mathfrak{J}(t) \in \pi$ (13.24). Hence $\mathfrak{J}(\mathfrak{M}^*) \subset \pi$. This completes the proof.

14. Details of the proof.

Here we want to supply the details omitted in section 13.

A. The definition (13.18) is independent of the representation. Let be $\bar{t}_1 = \bar{s}_1, \ldots, \bar{t}_r = \bar{s}_r$. We have to show that $\overline{ft_1 \ldots t_r} = \overline{fs_1 \ldots s_r}$. Using (13.12), (13.13) we find that $=t_1 s_1 \in \mathfrak{M}^*, \ldots, =t_r s_r \in \mathfrak{M}^*$. Now $=ft_1 \ldots t_r fs_1 \ldots s_r \in \mathfrak{M}^*$ by (10.7), (13.8). The proposition is now obtained by (13.12), (13.13).

B. The functions $\mathfrak{i}, \mathfrak{n}, \mathfrak{t}$, defined in (13.19), satisfy the condition (*), (**), (***) of section 3. This follows from the equivalence of the following propositions:

$\mathfrak{i}(\bar{s},\bar{t}) \in \Pi$	
$\tilde{\supset}(\bar{s},\bar{t}) \in \Pi$	(13.19)
$\overline{=st} \in \Pi$	(13.18)
$=st \in \mathfrak{M}^*$	(13.16)
$s \sim t$	(13.12)
$\bar{s} = \bar{t}$	(13.13),

$\mathfrak{n}(\bar{s}) \in \Pi$	
$\tilde{\neg}(\bar{s}) \in \Pi$	(13.19)
$\overline{\neg s} \in \Pi$	(13.18)
$\neg s \in \mathfrak{M}^*$	(13.16)
$s \notin \mathfrak{M}^*$	(13.9)
$\bar{s} \notin \Pi$	(13.16),

$$!(\bar{s},\bar{t}) \in \pi$$

$$\tilde{\wedge}(\bar{s},\bar{t}) \in \pi \qquad (13.19)$$

$$\overline{\wedge st} \in \pi \qquad (13.18)$$

$$\wedge st \in \mathfrak{M}^{*} \qquad (13.16)$$

$$s \in \mathfrak{M}^{*} \text{ and } t \in \mathfrak{M}^{*} \qquad (13.10)$$

$$\bar{s} \in \pi \text{ and } \bar{t} \in \pi \qquad (13.16).$$

C. Definition (13.21) is independent of the representatives.

Let be x_1, s_1 and x_2, s_2 two pairs of representatives for ρ. We have to show that $\overline{\varepsilon x_1 s_1} = \overline{\varepsilon x_2 s_2}$, i.e, by (13.13), (13.12), that $= \varepsilon x_1 s_1 \varepsilon x_2 s_2 \in \mathfrak{M}^{*}$.

From the fact that both x_1, s_1 and x_2, s_2 are pairs of representatives of ρ we get

$$(14.1) \qquad (s_1\frac{t}{x_1} \in \mathfrak{M}^{*} \quad \text{iff} \quad s_2\frac{t}{x_2} \in \mathfrak{M}^{*}) \qquad \text{for each } t\,.$$

Let be x a variable which is different from x_1, x_2 and does not occur free in s_1 or in s_2. Now by (12.11) and (13.8') we have

$$(14.2) \qquad (s_1\frac{x}{x_1}\frac{t}{x} \in \mathfrak{M}^{*} \quad \text{iff} \quad s_2\frac{x}{x_2}\frac{t}{x} \in \mathfrak{M}^{*}) \qquad \text{for each } t.$$

From (13.9), (13.10), (12.16) we obtain ($\alpha \in \mathfrak{M}^{*}$ iff $\beta \in \mathfrak{M}^{*}$) iff $\leftrightarrow\alpha\beta \in \mathfrak{M}^{*}$. Hence by (14.2)

$$(14.3) \qquad \leftrightarrow s_1\frac{x}{x_1}\frac{t}{x}\, s_2\frac{x}{x_2}\frac{t}{x} \in \mathfrak{M}^{*} \qquad \text{for each } t.$$

With section 11, B2 we get

$$(14.4) \qquad [\leftrightarrow s_1 \frac{x}{x_1} s_2 \frac{x}{x_2}] \frac{t}{x} \in \mathfrak{M}^* \qquad \text{for each } t.$$

Now for t we choose the term $\neg\leftrightarrow s_1 \frac{x}{x_1} s_2 \frac{x}{x_2}$ and obtain, using (12.18) :

$$(14.5) \qquad = \varepsilon x s_1 \frac{x}{x_1} \varepsilon x s_2 \frac{x}{x_2} \in \mathfrak{M}^* .$$

By the rule $(b^x_{x_1})$ we have $\vdash = \varepsilon x_1 s_1 \varepsilon x s_1 \frac{x}{x_1}$, and similarly $\vdash = \varepsilon x_2 s_2 \varepsilon x s_2 \frac{x}{x_2}$. Now from (14.5), (10.6), (10.5), (13.8) we get

$$(14.6) \qquad = \varepsilon x_1 s_1 \varepsilon x_2 s_2 \in \mathfrak{M}^* .$$

<u>D. The operator α defined in (13.21) is a choice operator.</u> We have to show that $\alpha(\rho) \in \rho$, if $\rho \neq o$. Let be x, s a pair of representatives for ρ. Let be $\bar{t} \in \rho$. According to (13.20) we have

$$(14.7) \qquad s\frac{t}{x} \in \mathfrak{M}^* .$$

Now the derivation

$$s \quad s\frac{\varepsilon x s}{x} \qquad (p_x)$$

$$s\frac{t}{x} \quad s\frac{\varepsilon x s}{x} \qquad (s^t_x), \text{ with } (11.9),$$

together with (14.7), (13.8), shows that $s\frac{\varepsilon xs}{x} \in \mathfrak{M}^*$. Hence, since x, s is a pair of representatives for ρ, we have $\overline{\varepsilon xs} \in \rho$, i.e. $\alpha(\rho) \in \rho$, which completes the proof.

E. Proof of (13.24). We show that $\mathfrak{J}(t) = \bar{t}$ by induction on the rank of t.

(14.8) If $t \equiv x$ is a variable, then by (13.23), (13.18)

$$\mathfrak{J}(x) = \tilde{x} = \bar{x}.$$

(14.9) If $t \equiv ft_1 \dots t_r$, we have

$$\begin{aligned} \mathfrak{J}(ft_1 \dots t_r) &= \mathfrak{J}(f)(\mathfrak{J}(t_1), \dots, \mathfrak{J}(t_r)) && (3.2') \\ &= \tilde{f}(\bar{t}_1, \dots, \bar{t}_r) && (13.23), \text{ind. hyp.} \\ &= \overline{ft_1 \dots t_r} && (13.18). \end{aligned}$$

(14.10) if $t \equiv \varepsilon xs$, we have

$$\begin{aligned} \mathfrak{J}(\varepsilon xs) &= \alpha(\{\bar{t}\colon \mathfrak{J}_x^{\bar{t}}(s) \in \pi\}) && (3.3') \\ &= \alpha(\{\bar{y}\colon \mathfrak{J}_x^{\bar{y}}(s) \in \pi\}) && (13.14) \\ &= \alpha(\{\bar{y}\colon \mathfrak{J}_x^{\mathfrak{J}(y)}(s) \in \pi\}) && (14.8) \\ &= \alpha(\{\bar{y}\colon \mathfrak{J}(s\tfrac{y}{x}) \in \pi\}) && (11.10) \\ &= \alpha(\{\bar{y}\colon \overline{s\tfrac{y}{x}} \in \pi\}) && \text{Ind. Hyp. (cf. (11.5))} \end{aligned}$$

$= \alpha(\{\bar{y}: s\frac{y}{x} \in \mathfrak{M}^*\})$ (13.16)

$= \alpha(\{\bar{t}: s\frac{t}{x} \in \mathfrak{M}^*\})$ see below (*)

$= \overline{\varepsilon xs}$ see below (**)

ad (*) : It is sufficient to show that $\{\bar{y}: s\frac{y}{x} \in \mathfrak{M}^*\} = \{\bar{t}: s\frac{t}{x} \in \mathfrak{M}^*\}$, i.e. that for every element $\mathfrak{r}$ of ω:

(14.11) $(Ex\ y)(\mathfrak{r} = \bar{y}$ and $s\frac{y}{x} \in \mathfrak{M}^*)$

iff $(Ex\ t)(\mathfrak{r} = \bar{t}$ and $s\frac{t}{x} \in \mathfrak{M}^*)$

The implication from left to right is trivial if t is chosen as y. From right to left we proceed as follows: Let be $\mathfrak{r} = \bar{t}$ and $s\frac{t}{x} \in \mathfrak{M}^*$. By (13.14) there is a variable y such that $\bar{t} = \bar{y}$. We only have to show that $s\frac{y}{x} \in \mathfrak{M}^*$. Since $\bar{t} = \bar{y}$ we have $=ty \in \mathfrak{M}^*$. By (12.12), (10.8), (10.3) we obtain $\vdash =ty\ s\frac{t}{x}\ s\frac{y}{x}$. Now we get the statement by (13.8).

ad (**) : The subclass ρ to which the choice operator α is applied, obviously is representable by the pair x, s. Hence the transition is justified by (13.21).

15. Completeness of restricted term calculus.

We want to show the second part of theorem 9.1 which has to do with restricted term calculus. What we have said in section 9, C and D, mutatis mutandis, also applies to restricted term calculus. According to (7.1) we obtain $\mathfrak{M} \vdash_1 \alpha$ from $\mathfrak{M} \vdash \alpha$. Hence we only have to show the analogue of the Theorem of satisfiability (13.1), where of course we have to replace the notion of consistency (which in (13.1) is related to the term calculus) by the analogous relation of consistency$_1$ (which is related to the restricted term calculus). The construction in section 13 also holds for restricted term calculus. It gives an interpretation, which satisfies $\mathfrak{M}$, over a semantical basis $\mathfrak{B} = \langle \omega, \pi, \mathcal{F}, \mathfrak{i}, \mathfrak{n}, \mathfrak{k}, \mathfrak{a} \rangle$, where in our case we only have to show that π has not more than one element. This can be shown as follows: In analogy to (13.8) we have

(15.1) If $\vdash_1 \alpha_1 \ldots \alpha_n \alpha$ and $\alpha_1, \ldots, \alpha_n \in \mathfrak{M}^*$, then $\alpha \in \mathfrak{M}^*$.

In analogy to (13.12) and (13.13) we get

(15.2) $\bar{s} = \bar{t}$ iff $=st \in \mathfrak{M}^*$.

π is defined as in (13.16):

(15.3) $\bar{s} \in \pi$ iff $s \in \mathfrak{M}^*$.

Let be $\bar{s} \in \pi$ and $\bar{t} \in \pi$. We have to show that $\bar{s} = \bar{t}$. With the help of the rule of restriction (R_x^y) and the rule of substitution we obtain

(15.4) $\vdash_1 s\ t\ =st$.

Since we have $\bar{s} \in \pi$ and $\bar{t} \in \pi$ we have $s \in \mathfrak{M}^*$ and $t \in \mathfrak{M}^*$ by (15.3). Now by (15.1) and (15.4) we get $=st \in \mathfrak{M}^*$, i.e. $\bar{s} = \bar{t}$ by (15.2).

Bibliography

[1] Hasenjaeger, G.: Eine Bemerkung zu Henkin's Beweis für die Vollständigkeit des Prädikatenkalküls der ersten Stufe, J. Symbolic Logic 18 (1953), 42-48.

[2] Henkin, L.: The completeness of the first order functional calculus. J. Symbolic Logic 14 (1949), 145 - 158.

[3] Henkin, L.: Completeness in the theory of types. J. Symbolic Logic 15 (1950), 81-91.

[4] Hermes, H.: Einführung in die mathematische Logik (^{2}1969).

[5] Hilbert, D. - P. Bernays: Grundlagen der Mathematik I (1934, ^{2}1968), II (1939).

[6] Rosser, B.: On the consistency of Quine's "New foundations for mathematical logic". J. Symbolic Logic 4 (1939), 15-24.

[7] Whitehead, A. N. - B. Russell: Principia Mathematica I, II, III (1910-13).

Also compare

[8] Carnap, R.: On the use of Hilbert's ε-operator in scientific theories. Essays on the Foundations of Mathematics dedicated to Prof. A. A. Fraenkel on his 70th anniversary. Magnes Press, The Hebrew University, Jerusalem 1961 (including a bibliography).

Index of Symbols

General Index

Offsetdruck: Julius Beltz, Weinheim/Bergstr.

_ecture Notes in Mathematics

isher erschienen/Already published

ɔl. 1: J. Wermer, Seminar über Funktionen-Algebren. IV, 30 Seiten. ·64. DM 3,80 / $ 1.10

ɔl. 2: A. Borel, Cohomologie des espaces localement compacts après. J. Leray. IV, 93 pages. 1964. DM 9,– / $ 2.60

ɔl. 3: J. F. Adams, Stable Homotopy Theory. Third edition. IV, 78 pages. ·69. DM 8,– / $ 2.20

ɔl. 4: M. Arkowitz and C. R. Curjel, Groups of Homotopy Classes. ıd. revised edition. IV, 36 pages. 1967. DM 4,80 / $ 1.40

ɔl. 5: J.-P. Serre, Cohomologie Galoisienne. Troisième édition. II, 214 pages. 1965. DM 18,– / $ 5.00

ɔl. 6: H. Hermes, Term Logic with Choise Operator. III, 55 pages. ·70. DM 6,– / $ 1.70

ɔl. 7: Ph. Tondeur, Introduction to Lie Groups and Transformation roups. Second edition. VIII, 176 pages. 1969. DM 14,– / $ 3.80

ɔl. 8: G. Fichera, Linear Elliptic Differential Systems and Eigenvalue ɔblems. IV, 176 pages. 1965. DM 13,50 / $ 3.80

l. 9: P. L. Ivănescu, Pseudo-Boolean Programming and Applications. 50 pages. 1965. DM 4,80 / $ 1.40

l. 10: H. Lüneburg, Die Suzukigruppen und ihre Geometrien. VI, Seiten. 1965. DM 8,– / $ 2.20

. 11: J.-P. Serre, Algèbre Locale. Multiplicités. Rédigé par P. Gabriel. conde édition. VIII, 192 pages. 1965. DM 12,– / $ 3.30

. 12: A. Dold, Halbexakte Homotopiefunktoren. II, 157 Seiten. 1966. 12,– / $ 3.30

. 13: E. Thomas, Seminar on Fiber Spaces. IV, 45 pages. 1966. 4,80 / $ 1.40

. 14: H. Werner, Vorlesung über Approximationstheorie. IV, 184 Sei- und 12 Seiten Anhang. 1966. DM 14,– / $ 3.90

. 15: F. Oort, Commutative Group Schemes. VI, 133 pages. 1966. 9,80 / $ 2.70

. 16: J. Pfanzagl and W. Pierlo, Compact Systems of Sets. IV, pages. 1966. DM 5,80 / $ 1.60

. 17: C. Müller, Spherical Harmonics. IV, 46 pages. 1966. 5,– / $ 1.40

18: H.-B. Brinkmann und D. Puppe, Kategorien und Funktoren. 107 Seiten, 1966. DM 8,– / $ 2.20

19: G. Stolzenberg, Volumes, Limits and Extensions of Analytic eties. IV, 45 pages. 1966. DM 5,40 / $ 1.50

20: R. Hartshorne, Residues and Duality. VIII, 423 pages. 1966. 20,– / $ 5.50

21: Seminar on Complex Multiplication. By A. Borel, S. Chowla, . Herz, K. Iwasawa, J.-P. Serre. IV, 102 pages. 1966. DM 8,– / $ 2.20

22: H. Bauer, Harmonische Räume und ihre Potentialtheorie. 75 Seiten. 1966. DM 14,– / $ 3.90

23: P. L. Ivănescu and S. Rudeanu, Pseudo-Boolean Methods for alent Programming. 120 pages. 1966. DM 10,– / $ 2.80

24: J. Lambek, Completions of Categories. IV, 69 pages. 1966. 6,80 / $ 1.90

25: R. Narasimhan, Introduction to the Theory of Analytic Spaces. 43 pages. 1966. DM 10,– / $ 2.80

26: P.-A. Meyer, Processus de Markov. IV, 190 pages. 1967. 15,– / $ 4.20

27: H. P. Künzi und S. T. Tan, Lineare Optimierung großer teme. VI, 121 Seiten. 1966. DM 12,– / $ 3.30

28: P. E. Conner and E. E. Floyd, The Relation of Cobordism to eories. VIII, 112 pages. 1966. DM 9,80 / $ 2.70

29: K. Chandrasekharan, Einführung in die Analytische Zahlen- rie. VI, 199 Seiten. 1966. DM 16,80 / $ 4.70

30: A. Frölicher and W. Bucher, Calculus in Vector Spaces without n. X, 146 pages. 1966. DM 12,– / $ 3.30

31: Symposium on Probability Methods in Analysis. Chairman. . Kappos. IV, 329 pages. 1967. DM 20,– / $ 5.50

32: M. André, Méthode Simpliciale en Algèbre Homologique et bre Commutative. IV, 122 pages. 1967. DM 12,– / $ 3.30

33: G. I. Targonski, Seminar on Functional Operators and tions. IV, 110 pages. 1967. DM 10,– / $ 2.80

34: G. E. Bredon, Equivariant Cohomology Theories. VI, 64 pages. . DM 6,80 / $ 1.90

35: N. P. Bhatia and G. P. Szegö, Dynamical Systems. Stability ry and Applications. VI, 416 pages. 1967. DM 24,– / $ 6.60

36: A. Borel, Topics in the Homology Theory of Fibre Bundles. 5 pages. 1967. DM 9,– / $ 2.50

Vol. 37: R. B. Jensen, Modelle der Mengenlehre. X, 176 Seiten. 1967. DM 14,– / $ 3.90

Vol. 38: R. Berger, R. Kiehl, E. Kunz und H.-J. Nastold, Differentialrechnung in der analytischen Geometrie IV, 134 Seiten. 1967 DM 12,– / $ 3.30

Vol. 39: Séminaire de Probabilités I. II, 189 pages. 1967. DM 14,– / $ 3.90

Vol. 40: J. Tits, Tabellen zu den einfachen Lie Gruppen und ihren Darstellungen. VI, 53 Seiten. 1967. DM 6.80 / $ 1.90

Vol. 41: A. Grothendieck, Local Cohomology. VI, 106 pages. 1967. DM 10,– / $ 2.80

Vol. 42: J. F. Berglund and K. H. Hofmann, Compact Semitopological Semigroups and Weakly Almost Periodic Functions. VI, 160 pages. 1967. DM 12,– / $ 3.30

Vol. 43: D. G. Quillen, Homotopical Algebra. VI, 157 pages. 1967. DM 14,– / $ 3.90

Vol. 44: K. Urbanik, Lectures on Prediction Theory. IV, 50 pages. 1967. DM 5,80 / $ 1.60

Vol. 45: A. Wilansky, Topics in Functional Analysis. VI, 102 pages. 1967. DM 9,60 / $ 2.70

Vol. 46: P. E. Conner, Seminar on Periodic Maps. IV, 116 pages. 1967. DM 10,60 / $ 3.00

Vol. 47: Reports of the Midwest Category Seminar I. IV, 181 pages. 1967. DM 14,80 / $ 4.10

Vol. 48: G. de Rham, S. Maumary et M. A. Kervaire, Torsion et Type Simple d'Homotopie. IV, 101 pages. 1967. DM 9,60 / $ 2.70

Vol. 49: C. Faith, Lectures on Injective Modules and Quotient Rings. XVI, 140 pages. 1967. DM 12,80 / $ 3.60

Vol. 50: L. Zalcman, Analytic Capacity and Rational Approximation. VI, 155 pages. 1968. DM 13.20 / $ 3.70

Vol. 51: Séminaire de Probabilités II. IV, 199 pages. 1968. DM 14,– / $ 3.90

Vol. 52: D. J. Simms, Lie Groups and Quantum Mechanics. IV, 90 pages. 1968. DM 8,– / $ 2.20

Vol. 53: J. Cerf, Sur les difféomorphismes de la sphère de dimension trois (Γ_4= O). XII, 133 pages. 1968. DM 12,– / $ 3.30

Vol. 54: G. Shimura, Automorphic Functions and Number Theory. VI, 69 pages. 1968. DM 8,– / $ 2.20

Vol. 55: D. Gromoll, W. Klingenberg und W. Meyer, Riemannsche Geometrie im Großen. VI, 287 Seiten. 1968. DM 20,– / $ 5.50

Vol. 56: K. Floret und J. Wloka, Einführung in die Theorie der lokalkonvexen Räume. VIII, 194 Seiten. 1968. DM 16,– / $ 4.40

Vol. 57: F. Hirzebruch und K. H. Mayer, O (n)-Mannigfaltigkeiten, exotische Sphären und Singularitäten. IV, 132 Seiten. 1968. DM 10,80 / $ 3.00

Vol. 58: Kuramochi Boundaries of Riemann Surfaces. IV, 102 pages. 1968. DM 9,60 / $ 2.70

Vol. 59: K. Jänich, Differenzierbare G-Mannigfaltigkeiten. VI, 89 Seiten. 1968. DM 8,– / $ 2.20

Vol. 60: Seminar on Differential Equations and Dynamical Systems. Edited by G. S. Jones. VI, 106 pages. 1968. DM 9,60 / $ 2.70

Vol. 61: Reports of the Midwest Category Seminar II. IV, 91 pages. 1968. DM 9,60 / $ 2.70

Vol. 62: Harish-Chandra, Automorphic Forms on Semisimple Lie Groups X, 138 pages. 1968. DM 14,– / $ 3.90

Vol. 63: F. Albrecht, Topics in Control Theory. IV, 65 pages. 1968. DM 6,80 / $ 1.90

Vol. 64: H. Berens, Interpolationsmethoden zur Behandlung von Approximationsprozessen auf Banachräumen. VI, 90 Seiten. 1968. DM 8,– / $ 2.20

Vol. 65: D. Kölzow, Differentiation von Maßen. XII, 102 Seiten. 1968. DM 8,– / $ 2.20

Vol. 66: D. Ferus, Totale Absolutkrümmung in Differentialgeometrie und -topologie. VI, 85 Seiten. 1968. DM 8,– / $ 2.20

Vol. 67: F. Kamber and P. Tondeur, Flat Manifolds. IV, 53 pages. 1968. DM 5,80 / $ 1.60

Vol. 68: N. Boboc et P. Mustață, Espaces harmoniques associés aux opérateurs différentiels linéaires du second ordre de type elliptique. VI, 95 pages. 1968. DM 8,60 / $ 2.40

Vol. 69: Seminar über Potentialtheorie. Herausgegeben von H. Bauer. VI, 180 Seiten. 1968. DM 14,80 / $ 4.10

Vol. 70: Proceedings of the Summer School in Logic. Edited by M. H. Löb. IV, 331 pages. 1968. DM 20,– / $ 5.50

Vol. 71: Séminaire Pierre Lelong (Analyse), Année 1967 – 1968. VI, 19 pages. 1968. DM 14,– / $ 3.90

Bitte wenden / Continued

Vol. 72: The Syntax and Semantics of Infinitary Languages. Edited by J. Barwise. IV, 268 pages. 1968. DM 18,– / $ 5.00

Vol. 73: P. E. Conner, Lectures on the Action of a Finite Group. IV, 123 pages. 1968. DM 10,– / $ 2.80

Vol. 74: A. Fröhlich, Formal Groups. IV, 140 pages. 1968. DM 12,– / $ 3.30

Vol. 75: G. Lumer, Algèbres de fonctions et espaces de Hardy. VI, 80 pages. 1968. DM 8,– / $ 2.20

Vol. 76: R. G. Swan, Algebraic K-Theory. IV, 262 pages. 1968. DM 18,– / $ 5.00

Vol. 77: P.-A. Meyer, Processus de Markov: la frontière de Martin. IV, 123 pages. 1968. DM 10,– / $ 2.80

Vol. 78: H. Herrlich, Topologische Reflexionen und Coreflexionen. XVI, 166 Seiten. 1968. DM 12,– / $ 3.30

Vol. 79: A. Grothendieck, Catégories Cofibrées Additives et Complexe Cotangent Relatif. IV, 167 pages. 1968. DM 12,– / $ 3.30

Vol. 80: Seminar on Triples and Categorical Homology Theory. Edited by B. Eckmann. IV, 398 pages. 1969. DM 20,– / $ 5.50

Vol. 81: J.-P. Eckmann et M. Guenin, Méthodes Algébriques en Mécanique Statistique. VI, 131 pages. 1969. DM 12,– / $ 3.30

Vol. 82: J. Wloka, Grundräume und verallgemeinerte Funktionen. VIII, 131 Seiten. 1969. DM 12,– / $ 3.30

Vol. 83: O. Zariski, An Introduction to the Theory of Algebraic Surfaces. IV, 100 pages. 1969. DM 8,– / $ 2.20

Vol. 84: H. Lüneburg, Transitive Erweiterungen endlicher Permutationsgruppen. IV, 119 Seiten. 1969. DM 10.– / $ 2.80

Vol. 85: P. Cartier et D. Foata, Problèmes combinatoires de commutation et réarrangements. IV, 88 pages. 1969. DM 8,– / $ 2.20

Vol. 86: Category Theory, Homology Theory and their Applications I. Edited by P. Hilton. VI, 216 pages. 1969. DM 16,– / $ 4.40

Vol. 87: M. Tierney, Categorical Constructions in Stable Homotopy Theory. IV, 65 pages. 1969. DM 6,– / $ 1.70

Vol. 88: Séminaire de Probabilités III. IV, 229 pages. 1969. DM 18,– / $ 5.00

Vol. 89: Probability and Information Theory. Edited by M. Behara, K. Krickeberg and J. Wolfowitz. IV, 256 pages. 1969. DM 18,– / $ 5.00

Vol. 90: N. P. Bhatia and O. Hajek, Local Semi-Dynamical Systems. II, 157 pages. 1969. DM 14,– / $ 3.90

Vol. 91: N. N. Janenko, Die Zwischenschrittmethode zur Lösung mehrdimensionaler Probleme der mathematischen Physik. VIII, 194 Seiten. 1969. DM 16,80 / $ 4.70

Vol. 92: Category Theory, Homology Theory and their Applications II. Edited by P. Hilton. V, 308 pages. 1969. DM 20,– / $ 5.50

Vol. 93: K. R. Parthasarathy, Multipliers on Locally Compact Groups. III, 54 pages. 1969. DM 5,60 / $ 1.60

Vol. 94: M. Machover and J. Hirschfeld, Lectures on Non-Standard Analysis. VI, 79 pages. 1969. DM 6,– / $ 1.70

Vol. 95: A. S. Troelstra, Principles of Intuitionism. II, 111 pages. 1969. DM 10,– / $ 2.80

Vol. 96: H.-B. Brinkmann und D. Puppe, Abelsche und exakte Kategorien, Korrespondenzen. V, 141 Seiten. 1969. DM 10,– / $ 2.80

Vol. 97: S. O. Chase and M. E. Sweedler, Hopf Algebras and Galois theory. II, 133 pages. 1969. DM 10,– / $ 2.80

Vol. 98: M. Heins, Hardy Classes on Riemann Surfaces. III, 106 pages. 1969. DM 10,– / $ 2.80

Vol. 99: Category Theory, Homology Theory and their Applications III. Edited by P. Hilton. IV, 489 pages. 1969. DM 24,– / $ 6.60

Vol. 100: M. Artin and B. Mazur, Etale Homotopy. II, 196 Seiten. 1969. DM 12,– / $ 3.30

Vol. 101: G. P. Szegö et G. Treccani, Semigruppi di Trasformazioni Multivoche. VI, 177 pages. 1969. DM 14,– / $ 3.90

Vol. 102: F. Stummel, Rand- und Eigenwertaufgaben in Sobolewschen Räumen. VIII, 386 Seiten. 1969. DM 20,– / $ 5.50

Vol. 103: Lectures in Modern Analysis and Applications I. Edited by C. T. Taam. VII, 162 pages. 1969. DM 12,– / $ 3.30

Vol. 104: G. H. Pimbley, Jr., Eigenfunction Branches of Nonlinear Operators and their Bifurcations. II, 128 pages. 1969. DM 10,– / $ 2.80

Vol. 105: R. Larsen, The Multiplier Problem. VII, 284 pages. 1969. DM 18,– / $ 5.00

Vol. 106: Reports of the Midwest Category Seminar III. Edited by S. Mac Lane. III, 247 pages. 1969. DM 16,– / $ 4.40

Vol. 107: A. Peyerimhoff, Lectures on Summability. III, 111 pages. 1969. DM 8,– / $ 2.20

Vol. 108: Algebraic K-Theory and its Geometric Applications. Edited by R. M. F. Moss and C. B. Thomas. IV, 86 pages. 1969. DM 6,– / $ 1.70

Vol. 109: Conference on the Numerical Solution of Differential Equations. Edited by J. Ll. Morris. VI, 275 pages. 1969. DM 18,– / $ 5.00

Vol. 110: The Many Facets of Graph Theory. Edited by G. Chartrand and S. F. Kapoor. VIII, 290 pages. 1969. DM 18,– / $ 5.00

Vol. 111: K. H. Mayer, Relationen zwischen charakteristischen Zahlen. III, 99 Seiten. 1969. DM 8,– / $ 2.20

Vol. 112: Colloquium on Methods of Optimization. Edited by N. N. Moiseev. IV, 293 pages. 1970. DM 18,– / $ 5.00

Vol. 113: R. Wille, Kongruenzklassengeometrien. III, 99 Seiten. 1970. DM 8,– / $ 2.20

Vol. 114: H. Jacquet and R. P. Langlands, Automorphic Forms on GL (2 VII, 548 pages. 1970. DM 24,– / $ 6.60

Vol. 115: K. H. Roggenkamp and V. Huber-Dyson, Lattices over Orders XIX, 290 pages. 1970. DM 18,– / $ 5.00

Vol. 116: Séminaire Pierre Lelong (Analyse) Année 1969. IV, 195 pages 1970. DM 14,– / $ 3.90

Vol. 117: Y. Meyer, Nombres de Pisot, Nombres de Salem et Analyse Harmonique. 63 pages. 1970. DM 6.– / $ 1.70

Vol. 118: Proceedings of the 15th Scandinavian Congress, Oslo 196… Edited by K. E. Aubert and W. Ljunggren. IV, 162 pages. 1970. DM 12,– / $ 3.30

Vol. 119: M. Raynaud, Faisceaux amples sur les schémas en groupe et les espaces homogènes. III, 219 pages. 1970. DM 14,– / $ 3.90

Vol. 120: D. Siefkes, Büchi's Monadic Second Order Successor Arithmetic. XII, 130 Seiten. 1970. DM 12,– / $ 3.30

Vol. 121: H. S. Bear, Lectures on Gleason Parts. III, 47 pages. 197… DM 6,–/$ 1.70

Vol. 122: H. Zieschang, E. Vogt und H.-D. Coldewey, Flächen und ebene diskontinuierliche Gruppen. VIII, 203 Seiten. 1970. DM 16,– $ 4.40

Vol. 123: A. V. Jategaonkar, Left Principal Ideal Rings. VI, 145 pages 1970. DM 12,– / $ 3.30

Vol. 124: Séminare de Probabilités IV. Edited by P. A. Meyer. IV, 2… pages. 1970. DM 20,– / $ 5.50

Vol. 125: Symposium on Automatic Demonstration. V, 310 pages. 197… DM 20,– / $ 5.50

Vol. 126: P. Schapira, Théorie des Hyperfonctions. XI, 157 pages. 19… DM 14,– / $ 3.90

Vol. 127: I. Stewart, Lie Algebras. IV, 97 pages. 1970. DM 10,– / $ 2…

Vol. 128: M. Takesaki, Tomita's Theory of Modular Hilbert Algebras and its Applications. II, 123 pages. 1970. DM 10,– / $ 2.80

Vol. 129: K. H. Hofmann, The Duality of Compact Semigroups a… C*- Bigebras. XII, 142 pages. 1970. DM 14,– / $ 3.90

Vol. 130: F. Lorenz, Quadratische Formen über Körpern. II, 77 Seit… 1970. DM 8,– / $ 2.20

Vol. 131: A Borel et al., Seminar on Algebraic Groups and Relat… Finite Groups. VII, 321 pages. 1970. DM 22,– / $ 6.10

Vol. 132: Symposium on Optimization. III, 348 pages. 1970. DM 22,– $ 6.10

Vol. 133: F. Topsøe, Topology and Measure. XIV, 79 pages. 197… DM 8,– / $ 2.20

Vol. 134: L. Smith, Lectures on the Eilenberg-Moore Spectral Sequen… VII, 142 pages. 1970. DM 14,– / $ 3.90

Vol. 135: W. Stoll, Value Distribution of Holomorphic Maps into Co… pact Complex Manifolds. II, 267 pages. 1970. DM 18,– / $

Vol. 136: M. Karoubi et al., Séminaire Heidelberg-Saarbrücken-St… buorg sur la K-Théorie. IV, 264 pages. 1970. DM 18,– / $ 5.00

Vol. 137: Reports of the Midwest Category Seminar IV. Edited… S. MacLane. III, 139 pages. 1970. DM 12,– / $ 3.30

Vol. 138: D. Foata et M. Schützenberger, Théorie Géométrique … Polynômes Eulériens. V, 94 pages. 1970. DM 10,– / $ 2.80

Vol. 139: A. Badrikian, Séminaire sur les Fonctions Aléatoires Li… aires et les Mesures Cylindriques. VII, 221 pages. 1970. DM 18, $ 5.00

Vol. 140: Lectures in Modern Analysis and Applications II. Edited… C. T. Taam. VI, 119 pages. 1970. DM 10,– / $ 2.80

Vol. 141: G. Jameson, Ordered Linear Spaces. XV, 194 pages. 19… DM 16,– / $ 4.40

Beschaffenheit der Manuskripte

Die Manuskripte werden photomechanisch vervielfältigt; sie müssen daher in sauberer Schreibmaschinenschrift geschrieben sein. Handschriftliche Formeln bitte nur mit schwarzer Tusche eintragen. Notwendige Korrekturen sind bei dem bereits geschriebenen Text entweder durch Überkleben des alten Textes vorzunehmen oder aber müssen die zu korrigierenden Stellen mit weißem Korrekturlack abgedeckt werden. Falls das Manuskript oder Teile desselben neu geschrieben werden müssen, ist der Verlag bereit, dem Autor bei Erscheinen seines Bandes einen angemessenen Betrag zu zahlen. Die Autoren erhalten 75 Freiexemplare.

Zur Erreichung eines möglichst optimalen Reproduktionsergebnisses ist es erwünscht, daß bei der vorgesehenen Verkleinerung der Manuskripte der Text auf einer Seite in der Breite möglichst 18 cm und in der Höhe 26,5 cm nicht überschreitet. Entsprechende Satzspiegelvordrucke werden vom Verlag gern auf Anforderung zur Verfügung gestellt.

Manuskripte, in englischer, deutscher oder französischer Sprache abgefaßt, nimmt Prof. Dr. A. Dold, Mathematisches Institut der Universität Heidelberg, Tiergartenstraße oder Prof. Dr. B. Eckmann, Eidgenössische Technische Hochschule, Zürich, entgegen.

Cette série a pour but de donner des informations rapides, de niveau élevé, sur des développements récents en mathématiques, aussi bien dans la recherche que dans l'enseignement supérieur. On prévoit de publier

1. des versions préliminaires de travaux originaux et de monographies
2. des cours spéciaux portant sur un domaine nouveau ou sur des aspects nouveaux de domaines classiques
3. des rapports de séminaires
4. des conférences faites à des congrès ou à des colloquiums

En outre il est prévu de publier dans cette série, si la demande le justifie, des rapports de séminaires et des cours multicopiés ailleurs mais déjà épuisés.

Dans l'intérêt d'une diffusion rapide, les contributions auront souvent un caractère provisoire; le cas échéant, les démonstrations ne seront données que dans les grandes lignes. Les travaux présentés pourront également paraître ailleurs. Une réserve suffisante d'exemplaires sera toujours disponible. En permettant aux personnes intéressées d'être informées plus rapidement, les éditeurs Springer espèrent, par cette série de »prépublications«, rendre d'appréciables services aux instituts de mathématiques. Les annonces dans les revues spécialisées, les inscriptions aux catalogues et les copyrights rendront plus facile aux bibliothèques la tâche de réunir une documentation complète.

Présentation des manuscrits

Les manuscrits, étant reproduits par procédé photomécanique, doivent être soigneusement dactylographiés. Il est recommandé d'écrire à l'encre de Chine noire les formules non dactylographiées. Les corrections nécessaires doivent être effectuées soit par collage du nouveau texte sur l'ancien soit en recouvrant les endroits à corriger par du verni correcteur blanc.

S'il s'avère nécessaire d'écrire de nouveau le manuscrit, soit complètement, soit en partie, la maison d'édition se déclare prête à verser à l'auteur, lors de la parution du volume, le montant des frais correspondants. Les auteurs recoivent 75 exemplaires gratuits.

Pour obtenir une reproduction optimale il est désirable que le texte dactylographié sur une page ne dépasse pas 26,5 cm en hauteur et 18 cm en largeur. Sur demande la maison d'édition met à la disposition des auteurs du papier spécialement préparé.

Les manuscrits en anglais, allemand ou français peuvent être adressés au Prof. Dr. A. Dold, Mathematisches Institut der Universität Heidelberg, Tiergartenstraße ou au Prof. Dr. B. Eckmann, Eidgenössische Technische Hochschule, Zürich.